Proceedings of the

Tenth Meeting on CPT and Lorentz Symmetry

Proceedings of the

Tenth Meeting on CPT and Lorentz Symmetry

Indiana University, Bloomington, USA 15 – 19 May 2025

Editor

Ralf Lehnert

Indiana University, USA

World Scientific

NEW JERSEY • LONDON • SINGAPORE • BEIJING • SHANGHAI • HONG KONG • TAIPEI • CHENNAI • TOKYO

Published by

World Scientific Publishing Co. Pte. Ltd.
5 Toh Tuck Link, Singapore 596224
USA office: 27 Warren Street, Suite 401-402, Hackensack, NJ 07601
UK office: 57 Shelton Street, Covent Garden, London WC2H 9HE

Library of Congress Control Number: 2026020315

British Library Cataloguing-in-Publication Data
A catalogue record for this book is available from the British Library.

PROCEEDINGS OF THE TENTH MEETING ON CPT AND LORENTZ SYMMETRY

ISBN 978-981-98-3346-7 (hardcover)
ISBN 978-981-98-3347-4 (ebook for institutions)
ISBN 978-981-98-3348-1 (ebook for individuals)

For any available supplementary material, please visit
https://www.worldscientific.com/worldscibooks/10.1142/14887#t=suppl

Desk Editors: Aanand Jayaraman/Carmen Teo Bin Jie

Typeset by Stallion Press
Email: enquiries@stallionpress.com

Preface

This proceedings volume collects contributions to the *Tenth Meeting on CPT and Lorentz Symmetry* (CPT'25), held at the Indiana University Physics Department in Bloomington from Thursday, May 15 to Sunday, May 19, 2025. Building on the series' longstanding format, CPT'25 showcased recent progress and emerging directions in experimental and theoretical efforts to probe possible violations of fundamental spacetime symmetries.

The chapters in this volume were solicited from conference participants who presented invited talks or posters. The meeting placed particular emphasis on experimental searches for deviations from CPT and Lorentz invariance and on the theoretical questions and frameworks that accompany such tests. On the experimental side, sessions featured astrophysical investigations involving neutrinos, photons, and gravitational waves; accelerator and storage-ring studies of neutral mesons, muons, quarks, and flavor-changing processes; laboratory tests of gravity; high-precision spectroscopy of ions, atoms, molecules, and exotic atoms; measurements of spin dynamics; comparisons of matter and antimatter; laser- and maser-based tests; neutron experiments; and studies using cavities, oscillators, resonators, as well as neutrino oscillations and propagation. Theoretical and phenomenological contributions addressed expected signals of CPT and Lorentz violation across particle physics, electromagnetism, and gravity; mechanisms and illustrative models for spacetime-symmetry breaking; developments in field theory, gravitation, and particle physics; and connections to condensed-matter systems.

The reports are arranged according to the schedule of lectures followed by the posters. CPT'25 benefited immensely from the continued support of Alan Kostelecký. I am also deeply appreciative of the many staff members and colleagues whose efforts were essential to the meeting, including Vy Munos, Yunhua Ding, River Govin, Connor Petway, Nathaniel Sherrill, and Albin Nilsson. Financial support from the Indiana University Center for Spacetime Symmetries (IUCSS), the Indiana University Office of the Vice Provost for Research, and the US Department of Energy is gratefully acknowledged.

Ralf Lehnert
May 2026

About the Editor

Ralf Lehnert obtained a Ph.D. in physics from Indiana University Bloomington in 2002 after completing his undergraduate education in physics at Dresden University of Technology. He is currently a Senior Research Scientist in Indiana University's Department of Physics and serves as the Associate Director of the Indiana University Center for Spacetime Symmetries.

Contents

Chapter 1

From Explicit Diffeomorphism Breaking to Spontaneous Unimodularization

Yuri Bonder* and Johas D. Morales

Instituto de Ciencias Nucleares,
Universidad Nacional Autónoma de México
Circuito Escolar s/n, Ciudad Universitaria,
04510, CdMx, México
**bonder@nucleares.unam.mx*

This contribution investigates modifications of General Relativity that allow or mimic energy nonconservation. We focus on Unimodular Gravity (UG), a theory that explicitly breaks diffeomorphism invariance, and show that it can lead to particle acceleration, opening the door to experimental constraints. We also argue that UG admits a well-posed initial value formulation, despite the presence of nondynamical structures. Finally, we introduce a simple model that spontaneously mimics UG, but in which a cosmological constant cannot arise as an integration constant.

1. Introduction

General Relativity (GR) is our most successful theory of gravity, validated across vast physical regimes.[1] Yet, GR faces profound challenges, especially its incompatibility with quantum mechanics. This points to a more fundamental theory predicting subtle deviations from GR, which can be sought by constructing new geometric gravitational models.

Energy conservation is a key principle in physics, but quantum mechanics does not strictly uphold it; for example, a Hamiltonian's expectation value changes discontinuously during energy measurement due to state collapse. Moreover, GR lacks a well-defined gravitational-energy notion, reinforcing the relaxation of this requirement for a fundamental gravity theory.

This raises the question: should theories bridging GR and quantum mechanics allow energy nonconservation? In GR, the conservation of the matter energy-momentum tensor, $T_{\mu\nu}$, is $\nabla^\mu T_{\mu\nu} = 0$, where ∇_μ is the covariant derivative associated with the metric $g_{\mu\nu}$, and the metric (inverse metric) is used to lower (raise) spacetime indices $\mu, \nu, \rho, \sigma, \ldots$. This contribution explores whether energy nonconservation can be incorporated into a geometric framework, with a focus on modified gravity theories that feature explicit or spontaneous breaking of invariance under diffeomorphisms.

2. Unimodular Gravity

Unimodular Gravity (UG) has been investigated by many authors.[2] The UG action, in four spacetime dimensions, can be written as

$$S = \int d^4x \, \frac{1}{2\kappa} \left[\sqrt{-g}\, R + \lambda \left(\sqrt{-g} - F\right)\right] + S_M(g, \{\psi\}), \tag{1}$$

where $\kappa = 8\pi G$, G is Newton's constant (units where the speed of light is one are used), $R = g^{\mu\nu} R_{\mu\nu}$ with $R_{\mu\nu}$ the Ricci tensor associated with $g_{\mu\nu}$, and $g = \det(g_{\mu\nu})$. Moreover, λ is a Lagrange multiplier enforcing the unimodular constraint, and F is a *nondynamical* scalar *density*. Also, S_M is the matter part of the action.

The equations of motion derived from this action include the unimodular constraint, $\sqrt{-g} = F$, the modified Einstein equation,

$$G_{\mu\nu} - \frac{1}{2}\lambda g_{\mu\nu} = \kappa T_{\mu\nu}, \tag{2}$$

where $G_{\mu\nu} = R_{\mu\nu} - \frac{1}{2} g_{\mu\nu} R$, and the matter field equations. Equation (2) resembles the Einstein equation of GR, with λ playing the role of a cosmological constant. However, there is no reason for λ to be constant.

Equation (2) can be manipulated by taking its trace, which yields

$$\lambda = -\frac{1}{2}\,(R + \kappa T), \tag{3}$$

where $T \equiv g^{\mu\nu} T_{\mu\nu}$. Inserting Eq. (3) into Eq. (2) leads to

$$G_{\mu\nu} + \frac{1}{4} R g_{\mu\nu} = \kappa \left(T_{\mu\nu} - \frac{1}{4} g_{\mu\nu} T\right), \tag{4}$$

which is the trace-free Einstein equation. This equation is interesting because, in a semiclassical approach, and under homogeneous and isotropic conditions, the vacuum expectation value of the energy-momentum tensor (derived from quantum fields) is proportional to the metric.[3] Hence, its trace-free part vanishes, ensuring its absence from Eq. (4).

Moreover, taking the divergence of Eq. (4) yields

$$\nabla_\nu R = \kappa \left(4 j_\nu - \nabla_\nu T\right), \tag{5}$$

where the Bianchi identity is used and $j_\mu \equiv \nabla^\nu T_{\mu\nu}$. Thus, under the additional assumption that $j_\mu = 0$, it implies

$$R + \kappa T = 4\Lambda, \tag{6}$$

where Λ is an arbitrary integration constant. When Eq. (6) is substituted into Eq. (4), it becomes

$$G_{\mu\nu} + \Lambda g_{\mu\nu} = \kappa T_{\mu\nu}, \tag{7}$$

which is the conventional Einstein equation with a cosmological constant. Notably, the cosmological constant here arises as an integration constant, decoupled from the matter-field vacuum state, thereby offering UG as a solution to the cosmological constant problem.[4] Also, when $j_\mu \neq 0$, the effective cosmological constant can be estimated from energy-nonconservation mechanisms, yielding observationally consistent values.[5]

It can be verified that the presence of the nondynamical scalar density F breaks diffeomorphism invariance, as is typically the case with nondynamical objects.[6] The variation under an infinitesimal diffeomorphism, generated by the vector field ξ^μ, of the gravitational action, $S_G = S - S_M$ yields

$$\delta S_G = \int d^4x \, \frac{1}{2\kappa} \left[-2\sqrt{-g} \left(G_{\mu\nu} - \frac{1}{2}\lambda g_{\mu\nu} \right) \nabla^\mu \xi^\nu + \xi^\mu \nabla_\mu \lambda \left(\sqrt{-g} - F\right) \right]. \tag{8}$$

The first and last terms cancel due to the Bianchi identity (after integrating by parts) and the unimodular constraint, respectively. Therefore, for the action to be invariant, it is necessary that the diffeomorphisms be along vector fields subject to $\nabla_\mu \xi^\mu = 0$, which is consistent with the interpretation of $F\, d^4x$ as a preferred spacetime volume element. Importantly, any divergence-free vector field can be written in terms of an *arbitrary* antisymmetric tensor field $\alpha_{\mu\nu}$ as

$$\xi^\mu = \epsilon^{\mu\nu\rho\sigma} \nabla_\nu \alpha_{\rho\sigma}, \tag{9}$$

where $\epsilon_{\mu\nu\rho\sigma}$ is the volume form, satisfying $\nabla_\lambda \epsilon_{\mu\nu\rho\sigma} = 0$.

Requiring the matter action to be invariant under diffeomorphisms generated by divergence-free vector fields leads to

$$\delta S_M = \int d^4x\sqrt{-g}\,\alpha_{\sigma\lambda}\,\epsilon^{\nu\rho\sigma\lambda}\,\nabla_\rho\nabla^\mu T_{\mu\nu} = 0, \tag{10}$$

where two integrations by parts are performed. Since this last equation is valid for all $\alpha_{\mu\nu}$, it implies

$$0 = \nabla_{[\nu}\nabla^\mu T_{\rho]\mu} = \nabla_{[\nu}j_{\rho]}, \tag{11}$$

where a pair of indices enclosed in square brackets is antisymmetrized with a factor of 1/2. A simple solution satisfying this condition is the usual conservation law: $j_\mu = 0$. Yet, this theory allows for more general matter behaviors. In the absence of topological obstructions, as assumed here, the most general solution is

$$j_\mu = -\nabla_\mu\Phi, \tag{12}$$

where Φ is an arbitrary scalar called the energy-loss potential. Note that the action does not determine the energy-loss potential explicitly; it merely allows for the possibility of a nontrivial one. In UG, the potential can be introduced as additional input. An energy-loss potential is proposed and tested in the next section.

2.1. *Empirical tests of Unimodular Gravity*

Tests involving particles and light have been extended to probe the energy loss allowed by UG.[7] By generalizing Papapetrou's method,[8] the equation of motion for a pointlike test particle described by a nonconserved energy-momentum tensor, $T^{\mu\nu}_{\text{part}}$ is

$$u^\nu\nabla_\nu u^\mu = J^\nu\,(\delta^\mu_\nu + u^\mu u_\nu), \tag{13}$$

where

$$J^\mu \equiv \frac{\lim\int_\Sigma d^3x\,\sqrt{h}\,\nabla_\nu T^{\mu\nu}_{\text{part}}}{\lim\int_\Sigma d^3x\,\sqrt{h}\,T^{tt}_{\text{part}}}, \tag{14}$$

and $d^3x\,\sqrt{h}$ is the volume element on a spacelike hypersurface Σ in Fermi coordinates associated with the particle's trajectory. The limits in the numerator and denominator arise because the method requires the particle to have finite extension during the derivation, with the point-particle limit taken only at the end.

Assuming spherical symmetry and that the energy-loss potential depends only on the radial coordinate, $\Phi = \Phi(r)$, the system effectively reduces to a single degree of freedom. The motion is then governed by the effective energy equation

$$E - \frac{1}{2} = \frac{1}{2}\dot{r}^2 + V_{\text{eff}}, \tag{15}$$

$$V_{\text{eff}} = \frac{1}{2g_{rr}}\left(\frac{\ell^2}{r^2} - \kappa\right) + \Phi(r) - \frac{1}{2}. \tag{16}$$

Here, ℓ is the conserved angular momentum associated with the Killing vector ∂_ϕ, while E is an integration constant. Note that there is no timelike Killing vector ∂_t. Moreover, $\kappa = -1, 0$ indicates whether the particle follows a timelike or null trajectory, respectively.

In addition, a Birkhoff-like theorem holds: if the traceless part of the Ricci tensor vanishes and spherical symmetry is imposed, the spacetime must be static.[7] The corresponding line element is

$$ds^2 = -\left(1 - \frac{2M}{r} + \tilde{\Lambda}r^2\right)dt^2 + \frac{dr^2}{1 - \frac{2M}{r} + \tilde{\Lambda}r^2} + r^2 d\Omega^2, \tag{17}$$

where M and $\tilde{\Lambda}$ are integration constants, interpreted as the gravitational mass of the source and an effective cosmological constant, respectively. The latter is neglected in the remainder of this work.

An energy-loss potential of the form

$$\Phi(r) = \frac{b_1^{\text{t}}}{r} + \frac{b_2^{\text{t}}}{r^2} + \frac{b_3^{\text{t}}}{r^3}, \tag{18}$$

is considered, where b_1^{t}, b_2^{t}, and b_3^{t} are model parameters chosen to reproduce the structure of the extrema in V_{eff}. The deflection angle, $\delta\phi$, for bound trajectories with semi-major axis a and eccentricity ϵ is approximated by

$$\delta\phi = \frac{6\pi M}{a(1-\epsilon^2)}\left(1 - \frac{b_2^{\text{t}}}{3M^2}\right) - \frac{6\pi b_3^{\text{t}}}{Ma^2(1-\epsilon^2)^2}. \tag{19}$$

Notably, $\delta\phi$ is insensitive to b_1^{t}. For the specific case of Mercury, the prefactor $6\pi M/[a(1-\epsilon^2)]$ corresponds to the well-known perihelion precession of approximately 43 arcsec per century, where M is the mass of the Sun.

Using PPN bounds,[9] the observed deflection angle is constrained by

$$\delta\phi^{\text{PPN}} = 43''\left(1 + \frac{2(\gamma-1) - (\beta-1)}{3}\right), \tag{20}$$

where

$$\gamma - 1 = (2.1 \pm 2.3) \times 10^{-5}, \tag{21}$$

$$\beta - 1 = (-4.1 \pm 7.8) \times 10^{-5}. \tag{22}$$

These constraints imply the bounds

$$-6.8 < (3.3 \times 10^4)\frac{b_2^{\iota}}{M^2} + 1.1\frac{b_3^{\iota}}{M^3} < 1.4. \tag{23}$$

Similar bounds hold for the case $\kappa = 0$, which is relevant for light bending.[7] This demonstrates that UG has observable empirical consequences. However, to establish UG as a viable theory, it must be predictive, an aspect explored in the next section.

2.2. *Initial value formulation*

Any viable physical theory must be predictive, or, in GR's argot, possess a well-posed initial value formulation. This requires identifying all constraints on the initial data and ensuring that the set of constraints are preserved under evolution. Furthermore, given initial data compatible with these constraints, the theory must yield a unique physical solution that depends continuously and causally on the initial data.[10]

The well-posedness of the initial value formulation in UG was analyzed.[11] It was shown that vacuum UG possesses a primary constraint, reminiscent of GR's momentum constraint. Requiring the preservation of this constraint under time evolution leads to a secondary constraint, analogous to the Hamiltonian constraint in GR with a cosmological constant. No further constraints are needed.

Moreover, using the BSSN formalism,[12] it was shown that even in the presence of a nontrivial energy-loss potential, the system remains hyperbolic. This, in turn, guarantees that the evolution is unique, continuous, and causal. Importantly, the unimodular condition, $F = \sqrt{-g}$, can be enforced through an appropriate choice of the lapse function. Finally, sufficient conditions on the matter action to maintain this well-posedness were identified, which are similar to those required in GR.

Still, these results suggest that a well-posed initial value problem exists because the nondynamical object in UG behaves essentially as pure gauge. If, instead, the nondynamical object had true physical content, it could potentially obstruct the initial value problem, as there would be no mechanism ensuring that its future values remain compatible with the dynamical evolution. We turn to study spontaneous unimodularization.

3. Spontaneous Unimodularization

Inspired by a question during the CPT'25 meeting, a theory replicating UG's features but spontaneously breaking diffeomorphism invariance is proposed, coining the term *spontaneous unimodularization*. For that purpose, one can propose the action

$$S = \int d^4x\sqrt{-g}\left[\frac{1}{2\kappa}R - \frac{1}{2}\nabla^\mu\left(\frac{F}{\sqrt{-g}}\right)\nabla_\mu\left(\frac{F}{\sqrt{-g}}\right) - V\right] + S_M(g,\psi), \tag{24}$$

where here $F > 0$ is a *dynamical* scalar density. Note that $F/\sqrt{-g}$ behaves like a scalar field and that the potential $V = V(F/\sqrt{-g})$ is chosen to drive F towards $\sqrt{-g}$. The variation of Eq. (24) leads to

$$\kappa T_{\mu\nu} = G_{\mu\nu} - \kappa\nabla_\mu\phi\nabla_\nu\phi + \kappa g_{\mu\nu}\left[\frac{1}{2}\nabla_\rho\phi\nabla^\rho\phi + V + \phi(\Box\phi - V')\right], \tag{25}$$

$$\Box\phi = V', \tag{26}$$

where the first equation is the Einstein equation and the second equation comes from the variation with respect to F. Here, we have abbreviated $\phi \equiv F/\sqrt{-g}$, $\Box\phi = \nabla_\mu\nabla^\mu\phi$, and the prime represents the derivative with respect to the function's argument. Remarkably, on shell, the last two terms in Eq. (25) cancel out and the dynamics is analogous to GR minimally coupled to a self-interacting Klein–Gordon field ϕ.

The strategy is to substitute Eq. (26) into Eq. (25) and follow the same steps as in UG. Taking the trace of the resulting equation yields

$$\kappa T = -R + \kappa\nabla_\mu\phi\nabla^\mu\phi + 4\kappa V. \tag{27}$$

Solving for V and substituting it back leads to

$$\kappa\left(T_{\mu\nu} - \frac{1}{4}g_{\mu\nu}T\right) = G_{\mu\nu} + \frac{1}{4}g_{\mu\nu}R - \kappa\left(\nabla_\mu\phi\nabla_\nu\phi - \frac{1}{4}g_{\mu\nu}\nabla_\rho\phi\nabla^\rho\phi\right), \tag{28}$$

which is traceless and resembles Eq. (4), up to the scalar field ϕ contribution. The divergence of Eq. (28) is

$$-\frac{\kappa}{4}\nabla_\nu T = \frac{1}{4}\nabla_\nu R - \kappa\left[\nabla^\mu(\nabla_\mu\phi\nabla_\nu\phi) - \frac{1}{4}\nabla_\nu(\nabla_\rho\phi\nabla^\rho\phi)\right], \tag{29}$$

where it is used that, since this model is diffeomorphism invariant, j_μ has to vanish. Note that the expression inside the brackets on the right-hand side can be manipulated as follows:

$$\begin{aligned}\nabla^\mu(\nabla_\mu\phi\nabla_\nu\phi)-\frac{1}{4}\nabla_\nu(\nabla_\rho\phi\nabla^\rho\phi)&=\Box\phi\nabla_\nu\phi+\frac{1}{2}\nabla^\mu\phi\nabla_\mu\nabla_\nu\phi\\&=V'\nabla_\nu\phi+\frac{1}{2}\nabla^\mu\phi\nabla_\mu\nabla_\nu\phi\\&=\nabla_\nu\left(V+\frac{1}{4}\nabla^\mu\phi\nabla_\mu\phi\right),\end{aligned}\tag{30}$$

where we have used Eq. (26) and properties of the covariant derivative. Consequently, Eq. (29) can be written as

$$0=\nabla_\nu\left(R+\kappa T-4\kappa V-\kappa\nabla^\mu\phi\nabla_\mu\phi\right),\tag{31}$$

whose solution, similarly to UG is

$$4\Lambda=R+\kappa T-4\kappa V-\kappa\nabla^\mu\phi\nabla_\mu\phi.\tag{32}$$

However, in stark contrast to UG, the separation between the trace and traceless parts of the Einstein equation is artificial in this case, and the trace equation is nontrivial. As a result, the integration constant Λ is not arbitrary but can be explicitly determined. In fact, it follows directly from Eq. (27) that $\Lambda=0$, implying that the mechanism of spontaneous unimodularization fails to generate a cosmological constant from an integration constant.

Still, it is possible to follow what is done in UG and introduce Eq. (32) into Eq. (28), which produces

$$\kappa\left(T_{\mu\nu}-\frac{1}{2}g_{\mu\nu}T\right)=R_{\mu\nu}-\kappa\nabla_\mu\phi\nabla_\nu\phi-g_{\mu\nu}\kappa V,\tag{33}$$

and whose trace reversed version is

$$\kappa T_{\mu\nu}=G_{\mu\nu}+\kappa g_{\mu\nu}V-\kappa\nabla_\mu\phi\nabla_\nu\phi+\frac{1}{2}g_{\mu\nu}\kappa\nabla^\mu\phi\nabla_\mu\phi.\tag{34}$$

This equation is the Einstein equation for the case where there is matter and an additional self-interacting field ϕ.

When V has a "deep" minimum and the field "lies" at the minimum, it acquires a value ϕ_0 such that $\nabla_\mu\phi_0$. Therefore, Eq. (35) becomes

$$\kappa T_{\mu\nu}=G_{\mu\nu}+g_{\mu\nu}\Lambda_{\text{eff}},\tag{35}$$

where $\Lambda_{\text{eff}}=\kappa V(\phi_0)$ acts as an effective cosmological constant. Notice that it is possible to use any potential with a minimum. However, the field may fluctuate around the minimum, producing interesting physics that is similar to what has been discussed extensively in the context of inflation.[13]

4. Conclusions

The introduction of nondynamical objects generally leads to unconventional physics, including the breaking of diffeomorphism invariance and energy nonconservation. In UG, this symmetry breaking opens the door for particular particle dynamics, which allows one to place bounds on possible mechanisms of energy loss. On the other hand, studying the initial value formulation of theories containing nondynamical objects is of limited value, since these objects have fixed "values" that cannot be predicted dynamically. However, UG represents a notable exception. In this theory, the nondynamical scalar density can be effectively absorbed into the lapse function. This feature provides a natural explanation for why UG constitutes such a mild deviation from GR, maintaining a well-posed initial value problem.

To explore whether UG phenomenology can be reproduced without invoking energy nonconservation, we considered a minimally coupled, self-interacting Klein–Gordon field ϕ, defined as the ratio of a dynamical scalar density F to $\sqrt{-g}$, and subject to a potential that induces spontaneous unimodularization. Upon taking the divergence of the traceless part of Einstein equation, we found that *no* integration constant emerges that could be reinstated into Einstein's equation. Nevertheless, when the field settles at the minimum of its potential, its value contributes to the effective cosmological constant; an effect reminiscent of the mechanism responsible for scalar-field-driven inflation.

Acknowledgments

We thank J.E. Herrera and A.M. Rubiol for their collaboration in Sec. 2. This work was supported by UNAM DGAPA-PAPIIT grant IN101724.

References

1. C.M. Will, *Theory and Experiment in Gravitational Physics,* Cambridge University Press, Cambridge, England, 2018.
2. See, e.g., J.L. Anderson and D. Finkelstein, *Am. J. Phys.* **39**, 901 (1971); W. Buchmuller and N. Dragon, *Phys. Lett. B* **207**, 292 (1988); W.G. Unruh, *Phys. Rev. D* **40**, 1048 (1989); M. Henneaux and C. Teitelboim, *Phys. Lett. B* **222**, 195 (1989); G.F.R. Ellis, H. van Elst, J. Murugan, and J.-P. Uzan, *Class. Quantum Grav.* **28**, 225007 (2011); Y. Bonder and C. Corral, *Phys. Rev. D* **97**, 084001 (2018); R. Carballo-Rubio, L.J. Garay, and G. García-Moreno, *Class. Quantum Grav.* **39**, 243001 (2022).
3. B.A. Juárez–Aubry, *Phys. Lett. B* **797**, 134912 (2019).
4. S. Weinberg, *Rev. Mod. Phys.* **61**, 1 (1989).

5. A. Perez and D. Sudarsky, *Phys. Rev. Lett.* **122**, 221302 (2019).
6. C. Corral and Y. Bonder, *Class. Quantum Grav.* **36**, 045002 (2019).
7. Y. Bonder, J.E. Herrera, and A.M. Rubiol, *Phys. Rev. D* **107**, 084032 (2023).
8. A. Papapetrou, *Proc. R. Soc. A* **209**, 248 (1951).
9. C.M. Will, *Living Rev. Relativity* **17**, 4 (2014); S.B. Lambert and C. Le Poncin-Lafitte, *Astron. Astrophys.* **529**, A70 (2011).
10. R.M. Wald, *General Relativity,* Chicago Univ. Press, 1984.
11. J.E. Herrera and Y. Bonder, *Phys. Rev. D* **109**, 104025 (2024).
12. M. Shibata and T. Nakamura, *Phys. Rev. D* **52**, 5428 (1995); T.W. Baumgarte and S.L. Shapiro, *Phys. Rev. D* **59**, 024007 (1998).
13. D. Baumann, "TASI Lectures on Inflation," arXiv:0907.5424 [hep-th].

Chapter 2

Bumblebee Gravity — Lessons from Perturbation Theory

N.A. Nilsson

Cosmology, Gravity and Astroparticle Physics Group,
Center for Theoretical Physics of the Universe,
Institute for Basic Science, Daejeon 34126, Korea
and
LTE, Observatoire de Paris, Université PSL,
CNRS, LNE, Sorbonne Université,
61 avenue de l'Observatoire, 75014 Paris, France

These proceedings summarize some recent efforts in understanding a class of vector-tensor theories known as *Bumblebee* models, which spontaneously break local Lorentz and diffeomorphism invariance. Using cosmological perturbation theory on an exact dS background, we find that for nonminimal coupling to gravity, the theory is strongly coupled unless a degeneracy condition is imposed, after which the model becomes a subset of generalized Proca theory, and we show that scalar perturbations become strongly coupled in the minimal-coupling limit. Moreover, we find a constraint on the bumblebee field from the speed of tensor modes on the order of 10^{-15}.

1. Introduction

The Standard Model of cosmology, with a hot Big Bang followed by an inflationary phase, eventually evolving to the dark-energy-dominated universe we live in today, is the prevailing cosmological model. Using the two main ingredients, general relativity and the cosmological principle, this model accurately describes the evolution from inflationary scales, where quantum effects dominate, to the formation and evolution of large-scale structure. That being said, the underlying theory of gravity, general relativity, is not without problems; for example, the cosmological constant problem sports a discrepancy of 55 orders of magnitude when comparing measurements and predictions from QFT,[1] and the Hubble parameter tension has reached particle-physics standards with a discrepancy of $> 5\sigma$.[2] These issues are deeply unsatisfactory as they keep us from achieving a truly elegant understanding and description of the Universe from primordial times until the present day. Whether from a fundamental misunderstanding of gravity at cosmological scales, hitherto hidden instrument systematics (such possibilities are discussed in the literature), or some other reason, there is now ample reason to study modifications of general relativity and/or one or more sectors of the Standard Model of particle physics.

The ultimate goal of physics can, in some sense, be said to be finding a "theory of everything," a single theory which gives accurate predictions at all scales, from the Big Bang to the present day and from cosmological scales to subatomic particles. Such a theory, sometimes called "quantum gravity," has been eluding scientists for over 50 years. There are many quantum-gravity candidate theories in the literature, for example string theory, loop quantum gravity, causal dynamical triangulation, Hořava–Lifshitz gravity, and more, with the latter three claiming to resolve the non-renormalizability problem of general relativity (and thus allowing for canonical quantization) whereas string theory attempts to unify all fundamental interactions. A feature which appears in several proposals to quantum gravity is that of *broken spacetime symmetries*, which was highlighted by Kostelecký & Samuel in Ref. 3, where it was shown that local Lorentz symmetry can be spontaneously broken in string field theory. As such, an EFT framework known as the Standard-Model Extension was developed[4] to help search for minute departures from exact local Lorentz, *CPT*, and diffeomorphism symmetry[a] and whilst several hints of violation have been found (see, for example, Refs. 5 and 6), it is not currently enough to claim a detection.

A popular vector-tensor theory which was first written down in 1989 in Refs. 7 and 8 and later found to be a vector subset of the Standard-Model Extension is known as the *Bumblebee model*, which incorporates spontaneous violation of local Lorentz symmetry (and therefore also

[a]For all current constraints, see the Data Tables.[9]

diffeomorphism symmetry). This model, thanks to its relative simplicity, has been the subject of a significant amount of study in the last decades,[b] especially in the context of compact objects such as Schwarzschild-like solutions,[10–12] rotating solutions,[13] and more. In cosmology, the literature is more scarce, with FLRW and AdS solutions at the background level,[14] cosmological tests with CMB data,[15] anisotropic cosmological solutions,[16] and more. We note the existence of a more general vector-tensor theory known as generalized Proca theory[17] which, although not strictly related to spacetime-symmetry breaking, contains several bumblebee models as subsets, and the literature on cosmology with generalized Proca is larger than that for the bumblebee model.[18–20] In these proceedings, we summarize some recent work on the existence of a map between bumblebee and generalized Proca, and identify a consistency condition necessary for a healthy bumblebee model. Finally, we show a constraint on the bumblebee background value from the speed of gravitational waves. These proceedings are based on the results in Ref. 21. We use $(-,+++)$ signature and units where $c = \hbar = 1$ and $G = 1/8\pi M_{\rm Pl}^2$.

2. The Model

Our starting point is the bumblebee action with nonminimal coupling to gravity as

$$\mathcal{L}_B \sim \frac{M_{\rm Pl}^2}{2} R + \xi B^\mu B^\nu R_{\mu\nu} + \sigma B_\mu B^\mu R - \frac{1}{4} B_{\mu\nu} B^{\mu\nu} - V_B\left(B^2\right), \tag{1}$$

where we note that in contrast to most of the bumblebee literature, the action is written such that the nonminimal couplings ξ and σ are dimensionless. Further, $M_{\rm Pl}$ is the Planck mass and V_B is the potential for the bumblebee field B^μ which spontaneously breaks the diffeomorphism (and hence local Lorentz) symmetry; moreover, since all operators are marginal, the only scale is introduced through the potential. Other terms can be considered in Eq. (1), but not all are equivalent, and can be rewritten using for example

$$R_{\mu\nu} B^\mu B^\nu = (\nabla_\mu B^\mu)^2 - \nabla_\mu B_\nu \nabla^\nu B^\mu + \text{total derivatives},$$

but the Lagrangian (1) is common in the literature and we choose to work with it.

---

[b] A search around the time of writing reveals over 180 papers in the last 10 years.

By varying (1) with respect to the inverse metric, we find the modified Einstein equations as

$$\begin{aligned}
&M_{\rm Pl}^2 G_{\mu\nu} + \xi\big[\nabla_\alpha\nabla_\beta\left(B^\alpha B^\beta\right) + \nabla_\alpha\nabla^\alpha\left(B_\mu B_\nu\right) - 2\nabla_\alpha\nabla_{(\mu}\left(B^\alpha B_{\nu)}\right)\\
&\quad - B^\alpha B^\beta R_{\alpha\beta}g_{\mu\nu} + 4B^\alpha B_{(\mu}R_{\nu)\alpha}\big] + 2\sigma\big[B^\alpha B_\alpha G_{\mu\nu}\\
&\quad + B_\mu B_\nu R + \left(g_{\mu\nu}\nabla_\beta\nabla^\beta - \nabla_\mu\nabla_\nu\right)B^\alpha B_\alpha\big] + g_{\mu\nu}V_B\\
&\quad - 2B_\mu B_\nu V_B' + g_{\mu\nu}\nabla_{[\alpha}B_{\beta]}\nabla^\alpha B^\beta - B_{\mu\alpha}B_\nu{}^\alpha = 0,
\end{aligned} \tag{2}$$

and the equation of motion for the bumblebee field is

$$\nabla_\alpha\nabla^\alpha B_\mu - \nabla^\alpha\nabla_\mu B_\alpha + 2\xi B^\alpha R_{\mu\alpha} + 2\sigma B_\mu R - 2B_\mu V_B' = 0. \tag{3}$$

3. Bumblebee Gravity on a Homogeneous and Isotropic Background

On a flat Friedmann–Lemaître–Robertson–Walker (FLRW) background with the metric in terms of cosmic time

$$ds^2 = -dt\otimes dt + a^2(t)\delta_{ij}dx^i\otimes dx^j, \tag{4}$$

where $a(t)$ is the cosmic scale factor, the Friedmann equations read

$$\begin{aligned}
&3\left(M_{\rm Pl}^2 + 6\sigma\bar{B}_0^2\right)H^2 + 6\bar{B}_0(\xi+2\sigma)\left(\bar{B}_0\dot{H} - H\dot{\bar{B}}_0\right) - V_B\\
&\quad - 2\bar{B}_0^2 V_B' = 0,\\
&\left(2(\xi+\sigma)\bar{B}_0^2 - M_{\rm Pl}^2\right)\left(3H^2 + 2\dot{H}\right) + 8H(\xi+\sigma)\bar{B}_0\dot{\bar{B}}_0\\
&\quad + 2(\xi+2\sigma)\left(\bar{B}_0\ddot{\bar{B}}_0 + \dot{\bar{B}}_0^2\right) + V_B = 0,
\end{aligned} \tag{5}$$

and the bumblebee equation of motion reduces to a constraint of the form

$$3(\xi+4\sigma)H^2 + 3(\xi+2\sigma)\dot{H} - V_B' = 0. \tag{6}$$

In the above equations, we have chosen a timelike ansatz for the bumblebee field, i.e.,

$$B_\mu \to \{\bar{B}_0(t), \vec{0}\},$$

which is a natural choice since the background metric is isotropic. From Eq. (6) we identify an interesting consequence of an expanding background: a *hallmark feature* of the bumblebee model, the vanishing of the first derivative of the potential at the background level, *no longer holds* on an expanding background, and we encounter a situation where the expansion of space, denoted by the Hubble parameter $H \equiv \dot{a}/a$, "kicks" the bumblebee field away from the potential minimum.

A second consequence of Eq. 6 is best illustrated by fixing the bumblebee potential V_B to be quartic, which is the most common choice in the bumblebee literature. We write it as

$$V_B\left(-\bar{B}_0^2\right) = \frac{\lambda}{2}(\bar{B}_0^2 - b^2)^2, \tag{7}$$

where λ is a dimensionless coupling and b is the "vacuum expectation value" triggered by the potential which on a Minkowski background induces $B_\mu \rightarrow \bar{B}_\mu = b_\mu$ and spontaneously breaks local Lorentz and diffeomorphism *particle* symmetry. This is not to be confused with *observer* Lorentz/diffeomorphism violation which leads to observer-dependent physics; this distinction has been treated at length elsewhere.[22,23] In a locally Lorentz frame, $b_a = e_a^\mu b_\mu \xrightarrow{\mathcal{P}} (b, \vec{0})$, b_a singles out a preferred direction and leads to violation of the equivalence principle.

With the potential defined as in Eq. (7) and restricting attention to exact dS, where the Hubble parameter is constant, we find

$$\bar{B}_0^2 = b^2 - \frac{3(\xi + 4\sigma)}{\lambda} H^2 = \text{const.}, \tag{8}$$

which is a striking result and has also been noticed elsewhere.[12,14] Since we may take a smooth limit of Eq. (6) from exact dS to Minkowski by setting the Hubble parameter to zero, the quantity $b = b_\mu b^\mu$ can be thought of as the Minkowski limit of the bumblebee background — it can be turned off without the bumblebee background contribution vanishing completely. In this sense, there exists a *screening mechanism* in the bumblebee model through which all spacetime symmetries may be restored at Solar-system scales.

4. Cosmological Perturbations of the Bumblebee Action — Existence of a Ghost Mode

We study cosmological perturbations of the model (1) around exact dS, and we decompose the bumblebee field as

$$B_\mu \rightarrow \{\bar{B}_0 + \epsilon\delta B_0,\ \epsilon\partial_i\delta B_s + \epsilon\delta B_i^{(T)}\}, \tag{9}$$

where we note the existence of two scalar modes and one divergenceless vector mode (not all of which are dynamical). We write the metric in ADM form as

$$ds^2 = -N(t)^2 dt^2 + a(t)^2 \gamma_{ij}(dx^i + N^i dt)(dx^j + N^j dt), \tag{10}$$

which is perturbed in spatially-flat gauge as

$$N(t) = 1 + \epsilon\alpha, \quad N^i = \frac{\epsilon}{a}\left(\mathcal{B}^i + \partial^i \beta\right), \quad \gamma_{ij} = \delta_{ij} + \epsilon h_{ij}, \tag{11}$$

where we have two scalar, one divergenceless vector, and two symmetric and trace-free tensor modes. We now perturb the action to second order in linear perturbations of scalars (α, β, δB_0, δB_s), vectors ($\mathcal{B}^i$, $(\delta B^\perp)^i$), and tensors (γ_{ij}). Not all of these degrees of freedom are dynamical and propagating, and we find that α can easily be integrated out of the quadratic action for scalar perturbations, after which we are left with three scalar degrees of freedom. After going to Fourier space, integrating by parts, and imposing the constraint and background equation, we can write, to quadratic order in scalar perturbations,

$$S_S^{(2)} = \frac{M_{\rm Pl}^2}{2} \int d^3x\, dt\, a^3 H^2 \left[\dot{\mathcal{V}}^\dagger \mathbf{K} \dot{\mathcal{V}} + k \dot{\mathcal{V}}^\dagger \mathbf{F} \mathcal{V} - \mathcal{V}^\dagger \mathbf{X} \mathcal{V}\right], \tag{12}$$

where $\mathbf{K}$ is the kinetic (Hessian) matrix, $\mathbf{F}$ is the friction matrix, and $\mathbf{X}$ contains the gradient and mass matrices. Also, we have defined the vector $\mathcal{V} \equiv \left(\beta,\ \delta B_0,\ (\delta B^\perp)_i\right)$. The kinetic matrix can be diagonalized and written

$$\mathbf{K} = \begin{pmatrix} \widetilde{K}_0 & 0 & 0 \\ 0 & \widetilde{K}_1 - \widetilde{K}_2 & 0 \\ 0 & 0 & \widetilde{K}_1 + \widetilde{K}_2 \end{pmatrix}, \tag{13}$$

which is a nondegenerate matrix of rank 3, and the model thus has a ghost, as no further constraint exists (we omit the exact form of $\{\widetilde{K}_0, \widetilde{K}_1, \widetilde{K}_2\}$ – they are given in Ref. 21). In order to cure the model, we need to identify a degeneracy condition in the non-minimal couplings ξ and σ, which turns out to be

$$\boxed{\sigma = -\frac{1}{2}\xi}, \tag{14}$$

at which point $\widetilde{K}_1$ and $\widetilde{K}_2$ both vanish, and we are left with one propagating scalar degree of freedom, as the rank of $\mathbf{K} \to 1$. We highlight that the degeneracy condition (14) is *independent of background and choice of potential*, as is known from GP; in other words

> *The bumblebee model* (1) *has a higher-derivative ghost instability unless the degeneracy condition* (14) *is imposed.*

With the degeneracy condition imposed, the bumblebee model becomes a subset of generalized Proca[c] with the identification

$$G_2 = -\frac{1}{4}B_{\mu\nu}B^{\mu\nu} - V_B\left(B^2\right), \quad G_4 = \frac{1}{2}\left(M_{\rm Pl}^2 - \xi B^2\right), \quad G_{4,X} = \xi, \tag{15}$$

at this point, all results obtained using generalized Proca theory apply, and the model is guaranteed to be stable and ghost free.[17]

After imposing the degeneracy condition (14), only the kinetic and gradient matrix survives and they both have rank one, meaning only one scalar degree of freedom is dynamical; the action can be written as

$$S_S^{(2)} = \frac{M_{\rm Pl}^2}{2}\int d^3x\,dt\,a^3\left[\mathcal{K}\delta\dot{B}_s^2 - \mathcal{G}\frac{k^2}{a^2}\delta B_s^2\right], \tag{16}$$

where we omit the kinetic and gradient coefficients $\mathcal{K}$ and $\mathcal{G}$, which can be found in Ref. 21. As we are interested in modes deep inside the Hubble horizon, we take the subhorizon limit $k \gg aH$ of the above action. In this limit, the kinetic coefficient $\mathcal{K}$ becomes independent of the nonminimal coupling ξ, and the gradient coefficient reads

$$\mathcal{G}_{k\gg aH} = \frac{8\lambda\xi\tilde{B}_0^2(\tilde{B}_0^2 - \tilde{b}^2)\left(1 + \xi(8\xi - 3)\tilde{B}_0^2\right)}{3(1 - 3\xi\tilde{B}_0^2)^2}, \tag{17}$$

which can be seen to have a global factor ξ. This in turn implies that the sound speed for scalar perturbations goes to zero in the minimal-coupling limit $\xi \to 0$ since

$$c_s^2|_{\xi\to 0} \equiv \left.\frac{\mathcal{G}}{\mathcal{K}}\right|_{\xi\to 0} = 0, \tag{18}$$

and the model thus has *strong-coupling issues* in the minimal-coupling limit.

4.1. *Tensor perturbations*

At second order in tensor perturbations, the action can be written as

$$S_T^{(2)} = \frac{M_{\rm Pl}^2}{8}\int d^3x\,dt\,a^3\left[\mathcal{K}_T\dot{h}_{jk}\dot{h}^{jk} - \frac{\mathcal{G}_T}{a^2}\partial_i h_{jk}\partial^i h^{jk}\right], \tag{19}$$

where the kinetic and gradient coefficients read

$$\begin{aligned}\mathcal{K}_T &= 1 - \xi\widetilde{B}_0^2,\\ \mathcal{G}_T &= 1 + \xi\widetilde{B}_0^2,\end{aligned} \tag{20}$$

[c]Although the bumblebee model predates generalized Proca by 35 years.

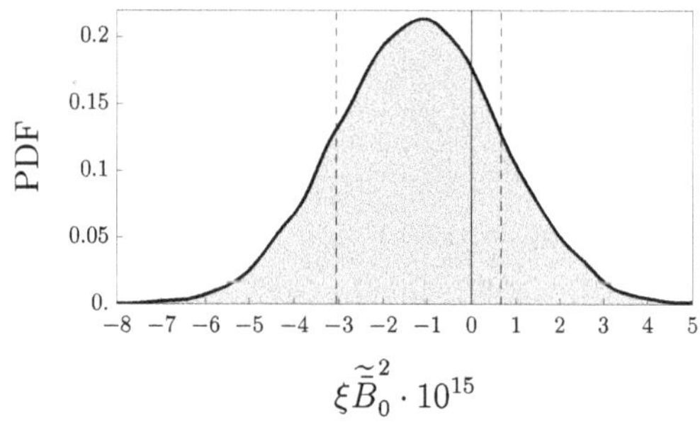

Fig. 1. Mock posterior constructed from 10^4 data points drawn from a Gaussian distribution using the current bound on the speed of gravitational waves. The dashed lines indicate the 1σ confidence levels.

from which we find the speed of tensor modes as

$$c_T^2 = \frac{2}{1 - \xi\tilde{\bar{B}}_0^2} - 1. \tag{21}$$

From the observation of an electromagnetic counterpart to the gravitational-wave event, it has been shown[24] that the speed of tensor modes must respect $-3 \cdot 10^{-15} < c_T - 1 < +7 \cdot 10^{-16}$. Knowing this, we assume that the error bars follow a Gaussian distribution and generate a posterior for $\xi\tilde{\bar{B}}_0^2$ by drawing 10^4 mock data points from this distribution, after which we find

$$\xi\tilde{\bar{B}}_0^2 = \left(-1.18^{+1.84}_{-1.87}\right) \cdot 10^{-15}, \tag{22}$$

the distribution for which can be seen in Fig. 1.

5. Discussion & Conclusions

In these proceedings, we have discussed cosmological perturbations of the bumblebee model on an exact dS background. By studying background evolution and perturbations, we conclude that

(1) with a quartic potential on an expanding background, the model has a screening mechanism, where Lorentz and diffeomorphism symmetries are restored in the Minkowski limit, as is shown in Eq. (8);

(2) the model has a higher-derivative ghost instability unless the degeneracy condition (14) is imposed. Once this is done, the model is stable at all orders and is a subset of generalized Proca theory with the map (15). This result is valid for *all backgrounds and all potentials*;

(3) the model is strongly coupled in the minimal-coupling limit $\xi \to 0$, at which point the model becomes unstable.

Acknowledgments

N.A.N. was financed by IBS under the project code IBS-R018-D3 and acknowledges support by PSL/Observatoire de Paris.

References

1. J. Martin, *Comptes Rendus Physique* **13**, 566 (2012).
2. E. Di Valentino *et al.* (CosmoVerse Network), *Phys. Dark Univ.* **49**, 101965 (2025).
3. V.A. Kostelecký and S. Samuel, *Phys. Rev. D* **39**, 683 (1989).
4. D. Colladay and V.A. Kostelecký, *Phys. Rev. D* **58**, 116002 (1998).
5. L. Haegel *et al.*, *Phys. Rev. D* **107**, 6, 6 (2023).
6. N.A. Nilsson and C. Le Poncin-Lafitte, *Phys. Rev. D* **109**, 015032 (2024).
7. V.A. Kostelecký and S. Samuel, *Phys. Rev. Lett.* **63**, 224 (1989).
8. V.A. Kostelecký and S. Samuel, *Phys. Rev. D* **40**, 1886 (1989).
9. V.A. Kostelecký and N. Russell, 2025 edition, *Data Tables for Lorentz and CPT Violation*, arXiv:0801.0287v18.
10. R. Casana *et al.*, *Phys. Rev. D* **97**, 104001 (2018).
11. R. Xu, D. Liang, and L. Shao, *Phys. Rev. D* **107**, 024011 (2023).
12. Q.G. Bailey, H.S. Murray, and D.T. Walter-Cardona, *Phys. Rev. D* **112**, 024069 (2025).
13. C. Ding *et al.*, *Eur. Phys. J. C* **80**, 178 (2020).
14. C. Reyes, M. Schreck, and A. Soto, *Phys. Rev. D* **106**, 023524 (2022).
15. R. Xu *et al.*, *Front. Phys.* (Beijing) **21**, 036201 (2026).
16. R.V. Maluf and J.C.S. Neves, *JCAP* **10**, 038.
17. L. Heisenberg, *JCAP* **05**, 015.
18. A. De Felice and A. Hell, *JHEP* **07**, 228 (2025).
19. A. De Felice *et al.*, *JCAP* **06**, 048 (2016).
20. J.C. Garnica *et al.*, *Ann. Phys.* **534**, 2100453 (2022).
21. N.A. Nilsson *et al.*, *On the Bumblebee – Generalised Proca Correspondence* (to be submitted).
22. V.A. Kostelecký, *Phys. Rev. D* **69**, 105009 (2004).
23. V.A. Kostelecký and Z. Li, *Phys. Rev. D* **103**, 024059 (2004).
24. B.P. Abbott *et al.*, *Astrophys. J. Lett.* **828**, L13 (2017).

Chapter 3

The QUPLAS Positronium Gravitation Experiment*

M. Giammarchi

Istituto Nazionale di Fisica Nucleare, Sezione di Milano, Via Celoria 16, 20133 Milano, Italy

QUPLAS (QUantum interferometry and gravitation with Positrons and LASers) aims at performing gravitational studies on a purely leptonic state: positronium (Ps), made of an electron and its antiparticle, the positron. These studies are relevant for testing the Einstein Equivalence Principle, a cornerstone of General Relativity, as well as CPT symmetry in particle physics.

1. Introduction

According to the most widely accepted cosmological scenario, when our Universe was in the *Electroweak Epoch* (between 10^{-35} and 10^{-12} s of cosmic time), an imbalance between particles and antiparticles manifested itself at the level of one part in 10^{10}. While the specific mechanism at the heart of this asymmetry is not fully understood, it is clear that a combination of cosmological and quantum conditions has to be met.

These conditions, put forward by A.D. Sakharov[1] in 1967, are as follows:

(1) CP and C violation.
(2) Baryon number violation.
(3) Departure from thermal equilibrium during the cosmological expansion.

*On behalf of the QUPLAS Collaboration.

As a result, at the end of the Electroweak Epoch a particle–antiparticle asymmetry was present, causing the complete prevalence of baryons over antibaryons at the beginning of the cosmic nucleosynthesis process, which took place during the first three minutes of cosmic time.

The possibility of CP violation is nicely incorporated into the framework of the Standard Model of particle physics by means of the Cabibbo Kobayashi–Maskawa matrix complex phase,[2] whose effect has been decisively proven in the hadronic sector. It appears, however, that the amount of CP violation accounted for by the Standard Model is largely insufficient to explain the amount of matter–antimatter asymmetry.[3]

In addition to this cosmic mystery, the main reason for studying antimatter systems is related to testing fundamental physical laws, such as Lorentz invariance, CPT symmetry, and the Einstein Equivalence Principle (EEP).[4]

These laws can be violated by effects coming from the Planck scale and motivated by string and quantum-gravity theories in a regime where the interplay between spacetime curvature and quantum effects cannot safely be neglected. Lorentz invariance, in particular, is called into play because of its role as a requisite of General Relativity and its link to CPT invariance, which can be broken only if Lorentz symmetry is also violated.[5] Therefore, the study of antimatter gravitation directly involves the role of the most general fundamental physical laws.

In this respect, the Standard-Model Extension (SME)[6,7] constitutes a phenomenological framework for testing Lorentz violation. The SME is an effective theory, starting from known fundamental physics (the Standard Model and General Relativity) and adding to it a series of Lorentz-violating spacetime operators, thereby allowing a systematic comparison with experiments and experimental limits. This makes possible a systematic study of Lorentz violations in many physical systems. In our case, we will focus on the gravitation of positronium, never measured before.[8]

2. Antihydrogen, Positronium, and the SME

Despite the great accuracy reached in testing the EEP for matter systems (down to one part in 10^{15}), very little is known of the gravitational properties of the antimatter sector, which is limited to the antihydrogen measurement conducted by the CERN ALPHA experiment, with ∼20% accuracy.[9]

While this measurement constitutes a milestone in the field, it refers to a system whose mass is mainly composed by the energy of the antiproton

QCD color field, leaving relatively little room for direct sensitivity to the behavior of Standard Model fundamental-constituent masses. Moreover, in the framework of the SME, this experimental result constrains only a combination of several Lorentz-violating parameters (see discussion in Ref. 10).

QUPLAS, on the other hand, focuses on positronium, the e^+e^- bound state, using only the (parallel-spin) ortho-Ps state, which has a lifetime of 142 ns. In addition to the obvious advantage of being composed almost entirely of fundamental fermion masses, its gravitational properties (in terms of a possible Lorentz violation) specifically pinpoint a single parameter in the SME.[10]

3. The QUPLAS Experiment

QUPLAS (QUantum interferometry and gravitation with Positrons and LASers) focuses on ortho-Ps gravitation, following the scheme of studying the propagation of a beam of Ps atoms in the Earth's gravitational field.

The general idea comes from our experience in antimatter interferometry[11,12] and from our gravitational measurements with atoms.[13] In QUPLAS, a beam of positronium atoms will be formed and then followed by a large-momentum-transfer (LMT) Mach–Zehnder interferometer to study its interaction with the Earth's gravitational field. The Ps beam will be flying at a velocity of roughly 10^7 m/s through the interferometer, in order to mitigate losses due to the short lifetime.

3.1. *The Ps beam*

In order to perform the measurement, it is necessary to provide an intense and well collimated Ps beam. The challenges of producing such a beam are complex, due to the instability and the neutrality of the system under consideration. The QUPLAS design calls for producing such a beam with a series of carefully designed steps,[15] schematically indicated in Fig. 1.

The production begins with a 9 MeV LINAC electron beam, which is converted to an e^+ beam with an energy peaked around 2 MeV. Its interaction in a thin target, covered with a Na monolayer, can generate the Ps^- ion. This state has a lifetime of only 0.479 ps,[14] but has the crucial feature of being charged, which allows its acceleration and collimation in a set of electrodes, reaching a potential of around 300 V.

After this step, a photodetachment takes place to remove the extra electron and finally form a well collimated ($\sim$ mrad) Ps beam. With this technique we are aiming at producing a $\sim 10^6$ Ps/s beam with a velocity

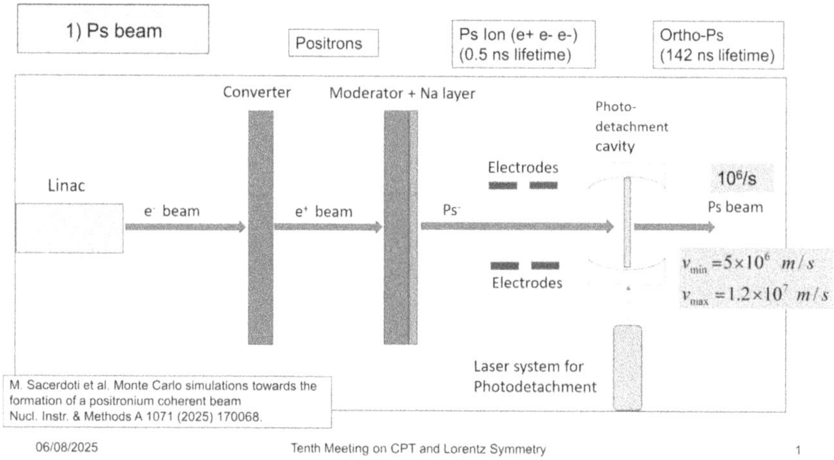

Fig. 1. Schematics of the formation of the Ps beam. After the generation of positrons by the initial electrons, the Ps^- ion is produced in a Na-decorated target. This ion is accelerated and collimated before photodetachment removes the extra electron. A set of velocities in the range indicated can be achieved.

in the range of 10^7 m/s. This velocity can be tuned (by a factor of two) to grant a better understanding of systematic effects.

3.2. *The Large-Momentum-Transfer (LMT) interferometer*

Once the Ps beam is formed, it will be subjected to the action of the LMT Mach–Zehnder interferometer, following the principles already used in Ref. 13. This is schematically presented in Fig. 2.

A preliminary excitation step will drive the Ps to reach the metastable 2^3S_1 (1.14 μs lifetime) state. Then, the rest of the interferometer will transfer a series of laser pulses in order to generate a phase between the two propagating states (2^3S_1 and 3^3P_2):

$$\Delta\phi = k_{\text{eff}} g T^2, \tag{1}$$

where k_{eff} is the total momentum transmitted to the states by the laser system, g is the gravitational acceleration, and T is the transit time in the interferometer. At the end of the apparatus, the $n = 3$ states will be ionized away and the phase will be measured by counting the number of $n = 2$ positronium atoms.

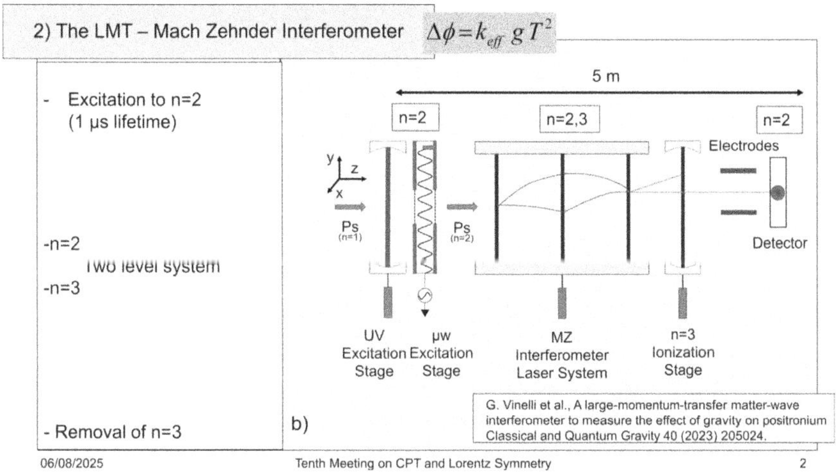

Fig. 2. Schematics of the working principle of the Mach–Zehnder interferometer. After the initial positronium excitation, the 5 m long interferometer will transfer a series of laser pulses to the system. Finally, the residual $n = 3$ states will be removed to allow the counting of only the $n = 2$ component.

The detailed measurement strategy involves the splitting of the Ps beam in two sections and the correlated study of the population in the two sections at the end of the interferometer, so that g will be extracted from the Lissajous-like contour as explained in detail in Ref. 17.

The full working principle and the detailed operations of the LMT interferometer are described in Ref. 16. Taking into account the predicted statistics and experimental configuration, the estimated accuracy of the g measurement will be 10%.

References

1. A.D. Sakharov, *JETP Lett.* **5**, 24 (1967).
2. M. Kobayashi and T. Maskawa, *Prog. Theor. Phys.* **49**, 652 (1973).
3. A.D. Dolgov, arXiv:hep-ph/9707419.
4. M. Giammarchi and G. Vinelli, *Universe* **8**, 123 (2022).
5. O.W. Greenberg, *Phys. Rev. Lett.* **89**, 231602 (2002).
6. D. Colladay and V.A. Kostelecký, *Phys. Rev. D* **55**, 6760 (1997).
7. D. Colladay and V.A. Kostelecký, *Phys. Rev. D* **58**, 116002 (1998).
8. *Data Tables for Lorentz and CPT Violation*, V.A. Kostelecký and N. Russell, 2025 edition, arXiv:0801.0287v18.
9. E.K. Anderson *et al.*, *Nature* **621**, 716 (2023).
10. V.A. Kostelecký and A. Vargas, *Phys. Rev. D* **92**, 056002 (2015).
11. S. Sala *et al.*, *Sci. Adv.* **5**, eaav7610 (2019).

12. M. Giammarchi, *Ann. Fond. Louis de Broglie* **48**, 1 (2024).
13. G. Rosi *et al.*, *Nature* **510**, 518 (2014).
14. Y. Nagashima, *Phys. Rep.* **545**, 95 (2014).
15. M. Sacerdoti *et al.*, *Nucl. Instrum. Methods A* **1071**, 170068 (2025).
16. G. Vinelli *et al.*, *Class. Quantum Grav.* **40**, 205024 (2023).
17. G. Rosi *et al.*, *Nat. Commun.* **8**, 15529 (2017).

Chapter 4

Progress in Lorentz and CPT Violation

V. Alan Kostelecký
Physics Department, Indiana University,
Bloomington, IN 47405, USA

This talk at the CPT'25 meeting presents an overview of some recent results in Lorentz and CPT violation.

1. Introduction

Lorentz transformations, which include spatial rotations and boosts, play a central role in our most successful description of nature at the fundamental level. In this description, the electromagnetic, weak, and strong interactions are encompassed by the Standard Model (SM), which is formulated on Minkowski spacetime and is invariant under global Lorentz transformations and translations. Gravitational effects are governed by General Relativity (GR), which involves Riemann spacetime and is invariant under local Lorentz transformations and diffeomorphisms.

An enticing idea that has received substantial attention in recent years is the notion that violations of Lorentz invariance and its concomitant CPT symmetry may be manifest in nature. These violations could emerge as residual effects from a unified theory of quantum physics and gravity at the Planck scale such as strings,[1] and the consequent small but observable signals continue to be the subject of numerous experimental searches.[2] This talk at the *Tenth Meeting on CPT and Lorentz Symmetry* (CPT'25) provides an overview of a few developments in the three years since the prior meeting in the series.

The treatment adopted here uses effective field theory[3] to describe prospective Lorentz- and CPT-violating effects in a realistic framework built on GR and the SM, known as the Standard-Model Extension (SME).[4–6] An outline of key concepts in this approach can be found in the proceedings of the prior meeting in this series.[7] The present contribution summarizes recent applications of this formalism to the geometry of Finsler spaces associated with Lorentz violation, the resolution of the concordance problem in Lorentz-violating effective field theories, the relativistic evolution of a spin in background fields, and the phenomenology of searches for flavor-changing Lorentz and CPT violation in charged leptons.

2. Finsler Geometry

The general Lagrange density describing a Dirac fermion propagating in a gravitational background can incorporate terms breaking local Lorentz and diffeomorphism symmetry. A term of this type involves a field operator transforming nontrivially under these symmetries and contracted with a coefficient governing the size of the physical effects.[5] When the coefficient is externally prescribed, any local Lorentz violation and diffeomorphism violation is nondynamical and hence explicit. This can result in a conflict between the spacetime and the particle dynamics if Riemann geometry is assumed,[5,8] producing unique experimental signals.[9] Adopting Finsler geometry instead is conjectured to resolve the conflict.[5] Finsler manifolds[10] provide a natural generalization of Riemann manifolds, in which the metric depends not only on location but also on direction in the tangent space. The conjecture relates the direction dependence in Finsler geometry to the prescribed coefficients in the Lagrange density.

In the extended Dirac theory, the analytic continuation of the classical trajectory of a relativistic wave packet in a fixed spacetime is indeed known to correspond to a geodesic in a Finsler geometry instead of a Riemann one.[11] Many of the specific examples of this correspondence studied to date are encompassed by the bipartite spaces, for which the Finsler norm F is defined to be the sum or difference of the conventional Riemann norm ρ and an additional seminorm σ built using a symmetric nonnegative 2-form $\mathbf{s}$. Among the set of bipartite spaces are the $\mathbf{a}$ spaces,[11] where $\mathbf{s}$ is constructed using the SME coefficient a_μ in the Dirac theory. The $\mathbf{a}$ spaces are related to Randers geometry, which underlies the Zermelo problem of navigation in a background wind.[12] Other bipartite spaces obtained from the Dirac theory include the **b** and **ab** spaces,[11] the **face** spaces,[13] certain **d**,[14] **g**,[14] **H**,[15] and **k**[16] spaces, and other special geometries.[17] The **b** spaces are of particular interest in that they arise from the SME coefficient b_μ governing the dominant CPT-odd spin-dependent modifications to the Dirac theory. The comparatively simple form of **s** for **b** spaces enables various results to be obtained,[11] including explicit expressions for geometric objects. Some other physical situations are known to involve **b** geometries, including a transversely magnetized chain and a bead sliding on a wire.[18]

A natural question is how to establish whether two given Finsler norms represent distinct geometries. Cartan discovered a symmetric torsion 3-tensor **C** that vanishes if and only if the Finsler manifold is a Riemann geometry.[19] Another symmetric 3-tensor **M**, known as the Matsumoto tensor,[20] vanishes when the Finsler manifold is a Randers geometry. The **a** spaces also have zero **M**.[11] Recently, a symmetric 3-tensor **S** that vanishes for bipartite spaces has been constructed.[21] The tensor **S** is expressed using the geometric quantities F and ρ and their derivatives, including the Cartan torsion **C** and the Finsler metric **g**. A symmetric 3-tensor **B** that is zero for **b** spaces and that can be specified in geometric terms has also been found.[21] These results provide support for the conjecture that all geometries arising from the extended Dirac theory can be characterized using appropriate symmetric 3-tensors. Identifying these tensors is an intriguing open problem that could lead to a classification of the possible geometries experienced by a Dirac fermion as it propagates in a fixed gravitational background in the presence of local Lorentz and diffeomorphism violations.

3. The Concordance Problem

The use of effective field theory to describe low-energy observables of Lorentz and CPT violation that could emerge from an underlying high-energy theory has parallels in condensed-matter physics. Prime examples of this are provided by Dirac and Weyl semimetals. For instance, a quasiparticle in a Weyl semimetal with band structure exhibiting isolated touching points has dynamics described by a generalized Dirac equation demonstrating experimentally measurable deviations from an emergent (3+1)-dimensional Lorentz invariance.[22] It is thus reasonable to hypothesize that the comprehensive SME framework can be adopted to investigate the properties of general quasiparticle excitations in the band structures of Dirac and Weyl materials, and conversely that established properties in these materials can be applied to open issues in the SME formalism.[23]

One longstanding issue in the SME approach has been the physical interpretation associated with large Lorentz violation. The difficulty occurs when the coefficients for Lorentz violation cannot be treated perturbatively in the chosen reference frame, or when a system with perturbative coefficients is viewed by a highly boosted observer. Under these circumstances, the description of the system can involve puzzling features such as negative energies. Finding a prescription that permits a consistent treatment of large coefficients for Lorentz violation, including in highly boosted observer frames, is called the concordance problem. Traditional analyses of physical Lorentz violation avoid the concordance problem by fiat, imposing the requirements that the coefficients are small and that boosts are constrained to concordant frames where perturbative Lorentz violation is maintained.[24] However, the parallels between the SME formalism and the treatment of Dirac and Weyl semimetals offer a novel opportunity to address the concordance problem.[25] In particular, no restriction on the magnitude of Lorentz violation appears in semimetals even for weak excitations, which suggests the concordance problem must be resolvable in the low-energy limit of at least some fundamental Lorentz-violating theories.

Following these ideas, a physical resolution of the concordance problem has recently been identified[25] for the theory of a Dirac fermion propagating in Minkowski spacetime with a nonzero SME coefficient b_μ for Lorentz violation.[4] In this model, the concordance problem is associated with the appearance of particles and antiparticles carrying negative energies. This feature occurs when the coefficient b_μ has components large compared to the fermion mass m in the chosen observer frame, or when components of b_μ that are small in one frame become large in the frame of a highly boosted observer. A resolution of the concordance problem in this case thus requires a physically viable understanding of the negative-energy particles and antiparticles.

Useful insight can be obtained by consideration of the properties of the corresponding Weyl semimetal, for which the Fermi surface is established via contact with the thermodynamic bath associated with the body of the semimetal. In particular, the ground state of the system is fixed by the Fermi surface, below which all the band energy levels are occupied. Note that this conventional description implicitly identifies the relevant reference frame as the rest frame of the bath, which is appropriate here because the bath rest frame is physically identical to the semimetal body.

These facts suggest using thermodynamic arguments to identify the physical ground state of the b_μ model via coupling to a generalized bath. Thermodynamic equilibrium then requires the physical vacuum to have zero Fermi energy, and so all negative-energy states must be filled. It follows that the ground state of the b_μ model has no positive-energy particles and antiparticles, as usual, but it does contain physical particles and antiparticles with negative energies. This situation is analogous to the physics in the Weyl semimetal, which includes filled physical levels with fermions and antifermions lying below the Fermi surface. In the b_μ model, the choice of bath and hence the relevant bath rest frame must be fixed by physical considerations. For prospective practical applications in fundamental physics, a natural choice of bath is the cosmological fluid in the Universe, which implies that the bath rest frame is the known cosmological rest frame associated with the cosmic microwave background. This hypothesis could in principle be experimentally tested if a physical species with large Lorentz violation is found.

Within this physical setup, the concordance problem for the fundamental particle model is resolved in the same way as in the Weyl semimetal. For example, in both cases the negative-energy particles and antiparticles in the ground state have no effect on stability because all levels are filled and hence no energy can be extracted from the ground state by transitions from positive to negative levels. The issue of large Lorentz boosts is also resolved because the bath fixes a physical rest frame, which implies that no observer boost can change the physics, instead at most changing the description of the physics along with the change of observer coordinates.

4. Extended Relativistic BMT Equation

The relativistic Bargmann–Michel–Telegdi (BMT) equation[26] provides a classical description of the behavior of a spin-$\frac{1}{2}$ particle in the presence of a homogeneous electromagnetic field. It encompasses the Heisenberg equation for the quantum spin operator while allowing for the Thomas precession associated with the particle acceleration, for the Larmor precession induced by an anomalous magnetic moment, and for the corresponding precession arising from an electric dipole moment. The BMT equation is relevant in realistic situations with charged fermions moving in homogeneous or near-homogeneous electromagnetic fields. For instance, it plays a key role in the analysis of data from storage-ring experiments seeking to measure the anomalous magnetic moment of the muon[27] or to discover an electric dipole moment of the muon or the proton.[28]

A natural question in this context is whether an extended relativistic BMT equation can be constructed that incorporates additional types of homogeneous background fields. These generically carry Lorentz indices, as a homogeneous background scalar field primarily acts to shift the fermion mass. Any homogeneous background with Lorentz indices can be identified with an SME coefficient for Lorentz violation that is spacetime constant in the relevant reference frame and that arises from explicit Lorentz violation. This line of reasoning indicates that the general extended relativistic BMT equation can be associated with the classical behavior of a spin-$\frac{1}{2}$ particle in the presence of explicit Lorentz violation. Indeed, the homogeneous electromagnetic field could also be viewed in this light, but for applications to searches for new physics it is convenient to maintain the distinction between effects from the electromagnetic field and those from other prospective backgrounds.

The Lagrange density for a Dirac spinor field interacting with a homogeneous electromagnetic field in the presence of Lorentz violation contains operators governing both propagation and interaction, and their explicit form has been established.[4,29,30] It is thus feasible in principle to construct the extended relativistic BMT equation for any homogeneous background. In a recent work,[31] this construction has been achieved for the dominant Lorentz-violating corrections, which arise from propagation effects involving operators of mass dimensions three and four and from electromagnetic couplings involving Lorentz-violating operators of mass dimensions five and six. The resulting extended relativistic BMT equation is relevant for searches for background physics beyond the SM in experiments with charged particles moving in electromagnetic fields. For example, it plays a role in searches for Lorentz violation using storage-ring experiments designed to seek electric dipole moments for the electron, proton, and muon.[31]

5. Charged-Lepton Flavor Violation

Flavor-changing physics in the minimal SM is driven by the weak interactions, leading to mixing of quark flavors and transitions from charged leptons to neutrinos. Beyond the minimal SM, neutrino oscillations reveal the existence of flavor-changing effects with lepton-number violation. Charged-lepton flavor-violating processes then must also occur, although they are one-loop suppressed with tiny branching ratios $\lesssim 10^{-54}$.[32] This implies that experiments seeking charged-lepton flavor-violating processes can provide sharp tests of new physics.[33]

Given this situation, it is natural to consider the prospects for experimental searches for Lorentz and CPT violation involving flavor changes in the charged leptons. Most investigations along these lines concern flavor changes during propagation rather than in interactions. For example, the dominant effects of Lorentz violation in neutrinos involve operators modifying neutrino mixing during propagation.[34] Similarly, the dominant Lorentz-violating effects in neutral-meson oscillations arise from modifications to the quark propagators.[35] However, Lorentz-violating interactions can also induce flavor-changing effects, notably in charged-lepton flavor-violating processes of interest here.

All flavor-changing operators of mass dimension $d \leq 6$ in the SME Lagrange density have been classified and enumerated.[30] Among these are Lorentz-violating operators involving electromagnetic interactions with $d = 5$, which contribute already at tree level to flavor-changing electromagnetic decays $\ell_A \to \ell_B + \gamma$ of a charged lepton ℓ_A of flavor A into a charged lepton ℓ_B of flavor $B \neq A$ and a photon.[36] Experimental constraints on the muon branching ratio of $\mathrm{BR}(\mu^+ \to e^+ + \gamma) \leq 4.2 \times 10^{-13}$ have been obtained by the Mu to Electron Gamma (MEG) collaboration at the Paul Scherrer Institute,[37] while limits for the tau of $\mathrm{BR}(\tau^\pm \to \mu^\pm + \gamma) \leq 4.4 \times 10^{-8}$ and $\mathrm{BR}(\tau^\pm \to e^\pm + \gamma) \leq 3.3 \times 10^{-8}$ have been placed by the *BABAR* collaboration at the Stanford Linear Accelerator.[38] These experimental results have been analyzed to yield limits expressed in the canonical Sun-centered frame[39] on coefficients for Lorentz violation governing electromagnetic flavor-changing interactions in the charged-lepton sector, lying in the range of parts in 10^{-13} GeV^{-1} to 10^{-9} GeV^{-1}.[36] Further analyses of these data allowing for sidereal and annual time variations and other planned experiments[40] offer interesting prospects for improved sensitivities.

Another golden channel for investigations of charged-lepton flavor-violating processes is the coherent conversion $\mu + N \to e + N$ of a muon into an electron in the presence of a nucleus N. In practice, the muon is captured in an atomic ground state, and the experimental signature is the monoenergetic spectrum arising from direct conversion. Tree-level contributions to this process are induced by Lorentz-violating operators involving electromagnetic interactions of mass dimension five or four-point interactions of leptons and quarks of mass dimension six in nuclear-mediated processes.[41] Using $^{197}_{79}$Au as the target nucleus, the SINDRUM II collaboration at the Paul Scherrer Institute has placed the tightest limit to date on the dimensionless ratio $R_{\mu e}$ of the conversion rate to the capture rate, $R_{\mu e} < 7 \times 10^{-13}$.[42] Forthcoming searches using $^{27}_{13}$Al as the target include one by the Muon-to-Electron Conversion (Mu2e) collaboration[43] at Fermilab, which aims to reach $R_{\mu e} \simeq 6.2 \times 10^{-16}$, and one by the Coherent Muon to Electron Transition (COMET) collaboration[44] at the Japan Proton Accelerator Complex (J-PARC), which expects to attain $R_{\mu e} \simeq 7 \times 10^{-15}$. The SINDRUM II data have recently been analyzed to yield limits at parts in 10^{-13} GeV^{-2} on coefficients for Lorentz violation controlling dimension-six flavor-changing four-point interactions of leptons and quarks, along with limits at parts in 10^{-12} GeV^{-1} on coefficients for Lorentz violation governing dimension-five flavor-changing electromagnetic interactions.[41] Future prospects include analyses of existing data using sidereal and annual variations, and improved experimental sensitivities are predicted in the forthcoming Mu2e and COMET experiments.[41]

Acknowledgments

This work was supported in part by US DoE grant DE-SC0010120 and by the Indiana University Center for Spacetime Symmetries.

References

1. V.A. Kostelecký and S. Samuel, *Phys. Rev. D* **39**, 683 (1989); V.A. Kostelecký and R. Potting, *Nucl. Phys. B* **359**, 545 (1991); *Phys. Rev. D* **51**, 3923 (1995).
2. V.A. Kostelecký and N. Russell, arXiv:0801.0287v18 (2025).
3. See, e.g., S. Weinberg, *Proc. Sci. CD* **09**, 001 (2009).
4. D. Colladay and V.A. Kostelecký, *Phys. Rev. D* **55**, 6760 (1997); *Phys. Rev. D* **58**, 116002 (1998).
5. V.A. Kostelecký, *Phys. Rev. D* **69**, 105009 (2004).
6. V.A. Kostelecký and Z. Li, *Phys. Rev. D* **103**, 024059 (2021).
7. V.A. Kostelecký, in R. Lehnert, ed., *Ninth Meeting on CPT and Lorentz Symmetry*, World Scientific, Singapore, 2023; arXiv:2210.09824.
8. R. Bluhm, *Phys. Rev. D* **91**, 065034 (2015); *Phys. Rev. D* **92**, 085015 (2015); R. Bluhm and A. Sehic, *Phys. Rev. D* **94**, 104034 (2016); R. Bluhm, H. Bossi, and Y. Wen, *Phys. Rev. D* **100**, 084022 (2019).
9. V.A. Kostelecký and Z. Li, *Phys. Rev. D* **104**, 044054 (2021).
10. P. Finsler, *Über Kurven und Flächen in allgemeinen Räumen*, University of Göttingen dissertation, 1918; Verlag Birkhäuser, Basel, Switzerland, 1951.
11. V.A. Kostelecký, *Phys. Lett. B* **701**, 137 (2011).
12. E. Zermelo, Z. Angew. Math. Mech. **11**, 114 (1931); Z. Shen, Canad. J. Math. **55**, 112 (2003); D. Bao and C. Robles, in D. Bao *et al.*, eds., *A Sampler of Riemann-Finsler Geometry*, Cambridge University Press, Cambridge, 2004.
13. V.A. Kostelecký and N. Russell, *Phys. Lett. B* **693**, 443 (2010).

14. J.A.A.S. Reis and M. Schreck, *Phys. Rev. D* **97**, 065019 (2018).
15. V.A. Kostelecký, N. Russell, and R. Tso, *Phys. Lett. B* **716**, 470 (2012).
16. B.R. Edwards and V.A. Kostelecký, *Phys. Lett. B* **786**, 319 (2018).
17. J.E.G. Silva, R.V. Maluf, and C.A.S. Almeida, *Phys. Lett. B* **798**, 135009 (2019).
18. J. Foster and R. Lehnert, *Phys. Lett. B* **746**, 164 (2015).
19. E. Cartan, *Les Espaces de Finsler*, Hermann, Paris, France, 1934; A. Deicke, *Arch. Math.* **4**, 45 (1953).
20. M. Matsumoto, *Tensor*, NS **24**, 29 (1972).
21. J.F. Davis, B.R. Edwards, and V.A. Kostelecký, arXiv:2508.21744.
22. B.Q. Lv *et al.*, *Phys. Rev. X* **5**, 031013 (2015); A. Tamai *et al.*, *Phys. Rev. X* **6**, 031021 (2016); B. Yan and C. Felser, *Annu. Rev. Condens. Matter Phys.* **8**, 337 (2017); N.P. Armitage, E.J. Mele, and A. Vishwanath, *Rev. Mod. Phys.* **90**, 015001 (2018); H. Gao *et al.*, *Ann. Rev. Mater. Res.* **49**, 153 (2019).
23. V.A. Kostelecký, R. Lehnert, N. McGinnis, M. Schreck, and B. Seradjeh, *Phys. Rev. Res.* **4**, 023106 (2022); A. Gómez, A. Martín-Ruiz, and L. Urrutia, *Phys. Lett. B* **829**, 137043 (2022).
24. V.A. Kostelecký and R. Lehnert, *Phys. Rev. D* **63**, 065008 (2001).
25. V.A. Kostelecký, R. Lehnert, M. Schreck, and B. Seradjeh, *New J. Phys.* **27**, 073901 (2025); *Phys. Lett. B* **865**, 139414 (2025); *Phys. Rev. D*, in press (arXiv:2512.01156).
26. V. Bargmann, L. Michel, and V.L. Telegdi, *Phys. Rev. Lett.* **2**, 435 (1959).
27. J. Bailey *et al.*, Nucl. Phys. B **150**, 1 (1979); G.W. Bennett *et al.*, *Phys. Rev. D* **73**, 072003 (2006); D.P. Aguillard *et al.*, *Phys. Rev. Lett.* **131**, 161802 (2023).
28. G.W. Bennett *et al.*, *Phys. Rev. D* **80**, 052008 (2009); J. Pretz, *Hyperfine Interact* **214**, 111 (2013); V. Anastassopoulos *et al.*, *Rev. Sci. Instrum.* **87** 115116, (2016); F. Rathmann and N.N. Nikolaev, *PoS* **SPIN2018**, 004 (2019); Y. Takeuchi, *JPS Conf. Proc.* **37**, 020603 (2022); J. Alexander *et al.*, arXiv:2205.00830; P. Schmidt-Wellenberg *et al.*, *PoS* **Muon4Future2023**, 018 (2023); H. Ströher *et al.*, *Particles* **6**, 385 (2023).
29. V.A. Kostelecký and M. Mewes, *Phys. Rev. D* **88**, 096006 (2013); Y. Ding and V.A. Kostelecký, *Phys. Rev. D* **94**, 056008 (2016).
30. V.A. Kostelecký and Z. Li, *Phys. Rev. D* **99**, 056016 (2019).
31. Y. Ding, V.A. Kostelecký, and A.J. Vargas, *Phys. Lett. B* **867**, 139604 (2025).
32. W.J. Marciano and A.I. Sanda, *Phys. Lett. B* **67**, 303 (1977); S.M. Bilenky, S.T. Petcov, and B. Pontecorvo, *Phys. Lett. B* **67**, 309 (1977); B.W. Lee, S. Pakvasa, R.E. Schrock, and H. Sugawara, *Phys. Rev. Lett.* **38**, 937 (1977).
33. R. Barbieri and L.J. Hall, *Phys. Lett. B* **338**, 212 (1994); R.H. Bernstein and P.S. Cooper, *Phys. Rept.* **532**, 27 (2013); L. Calibbi and G. Signorelli, *Riv. Nuovo Cim.* **41**, 71 (2018).
34. V.A. Kostelecký and M. Mewes, *Phys. Rev. D* **85**, 096005 (2012).
35. V.A. Kostelecký, *Phys. Rev. Lett.* **80**, 1818 (1998).
36. V.A. Kostelecký, E. Passemar, and N. Sherrill, *Phys. Rev. D* **106**, 076016 (2022).

37. A.M. Baldini *et al.*, *Eur. Phys. J. C* **76**, 434 (2016).
38. B. Aubert *et al.*, *Phys. Rev. Lett.* **104**, 021802 (2010).
39. R. Bluhm *et al.*, *Phys. Rev. D* **68**, 125008 (2003); *Phys. Rev. Lett.* **88**, 090801 (2002); V.A. Kostelecký and M. Mewes, *Phys. Rev. D* **66**, 056005 (2002).
40. A.M. Baldini *et al.*, *Eur. Phys. J. C* **78**, 380 (2018); E. Kou *et al.*, *PTEP* **2019**, 123C01 (2019).
41. V.A. Kostelecký, W. McNulty, E. Passemar, and N. Sherrill, *Phys. Lett. B* **868**, 139755 (2025).
42. W.H. Bertl *et al.*, *Eur. Phys. J. C* **47**, 337 (2006); A. van der Schaaf, *SciPost Phys. Proc.* **5**, 008 (2021).
43. L. Bartoszek *et al.*, arXiv:1501.05241; K. Byrum *et al.*, *Universe* **2023**, 54 (2023).
44. R. Abramishvili *et al.*, *PTEP* **2020**, 033C01 (2020).

Chapter 5

Tests of Lorentz Symmetry Using X-ray Polarimetry

F. Kislat

Department of Physics & Astronomy and Space Science Center,
University of New Hampshire,
Durham, NH 03824, USA

Lorentz symmetry is the fundamental symmetry of Einstein's theory of Special Relativity and has been tested to great precision. Nevertheless, the possibility remains that it is violated at the Planck scale, as predicted by some theories of quantum gravity. While the Planck scale is not directly accessible to experiments, minute residual deviations from Lorentz symmetry at attainable energies may be observable. The polarization of light from astrophysical sources is a particularly powerful probe because tiny differences accumulate as light travels over astrophysical distances, and polarization is sensitive to light travel time differences between polarization modes on the order of the oscillation period of the electromagnetic wave. Here, we report on new constraints on Lorentz invariance violation derived from X-ray polarization measurements of active galactic nuclei. The new constraints, presented in the framework of the Standard-Model Extension, improve on our previous work, which used optical polarization measurements, by four orders of magnitude.

1. Introduction

Although Lorentz invariance has been tested to great precision,[1] some theories of quantum gravity indicate that the symmetry may be broken at the Planck scale.[2] Residual effects may exist at attainable energies, but numerous experiments have established that, if present, these effects must be tiny.

As a consequence of Lorentz symmetry, the speed of light is the same in all reference frames, invariant under boosts and rotations. This implies that the speed of light must be independent of wavelength (or photon energy), propagation direction, and polarization.

When photons propagate over astronomical distances, even minuscule deviations from the speed of light, c, will accumulate and may result in observable effects. As a result, some of the strongest constraints on Lorentz invariance violations are due to astrophysical and cosmological observations.[1] The greatest challenge of such tests is that the source of the observed light cannot be controlled and details of the emission are generally unknown. Nevertheless, robust constraints on Lorentz-violating theories can be derived from observations by making a few conservative assumptions, illustrated most easily with the help of two examples.

Explosive events such as gamma-ray bursts (GRBs) and supernova explosions result in the emission of large amounts of radiation on timescales as short as seconds. While the absolute propagation time cannot be measured, it is possible to compare the arrival time of photons at different energies, enabling a test of the photon dispersion relation.[3] For a typical GRB, arrival time differences can be constrained to $\mathcal{O}(1\,\mathrm{s})$. In general, such measurements are limited by photon statistics, not the timing precision of the instrument. If the GRB is at redshift $z = 1$ (light travel time $\sim 7.7 \times 10^{12}\,\mathrm{a}$), this amounts to a sensitivity to differences in propagation velocity of $\mathcal{O}(10^{-20})$. Of course, this estimate implies the assumption that light of all wavelengths is emitted at the same time. That need not be true, but conservative tests can be constructed in such a way that reliable upper limits on velocity differences can be derived except in the unlikely case that astrophysical effects and Lorentz-invariance-violating effects cancel.

Even tighter constraints can be obtained from polarization measurements.[4–7] Light travel time differences between different polarization modes on the order of the oscillation period of the wave, ν^{-1}, will result in a measurable change of the polarization angle compared to the source. The absence of such a change then results in a constraint on birefringence that is a factor $\mathcal{O}(\nu^{-1})$ stronger than time-of-flight constraints from the same source. Although the polarization at the source is unknown, any energy dependence of such an effect will lead to an energy-dependent observed polarization even if the polarization at the source does not depend on energy. The assumption of energy-independent polarization at the source can often be motivated if the emission mechanism is the same across the observational energy range. Even if this is not the case, a non-detection of energy dependence can be used to derive constraints on birefringence under the assumption that it is very unlikely that astrophysical and Lorentz-violating effects will cancel.

The Standard-Model Extension (SME) is an effective field theory framework to describe Lorentz- and CPT-invariance-violating effects in theories beyond the Standard Model.[6,8] The effect of each of the higher-order operators is described by a set of coefficients, which can be ordered by the mass dimension d of the corresponding operator. Working in this framework has the advantage that it enables a direct comparison of results obtained from different experimental approaches and to translate those results into constraints on the underlying theories.

Previously, we have used optical polarimetry of active galactic nuclei (AGN) and GRB afterglows to derive some of the strongest constraints on the $d = 4, 5, 6$ coefficients of the SME photon sector.[7,9] Here, we extend this approach to include X-ray polarimetry results from the Imaging X-ray Polarimetry Explorer[10] mission (IXPE) and derive new constraints on the $d = 5$ isotropic model.

The remainder of the paper is structured as follows: We describe our methodology in Section 2, present the results in Section 3, and finally give a summary and outlook towards future work in Section 4.

2. Methods

To explain our approach, we begin with a brief review of the relevant theory following Ref. 6. In the SME, Lorentz-violating operators result in a modified photon dispersion relation,

$$E \simeq \left(1 - \varsigma^0 \pm \sqrt{(\varsigma^1)^2 + (\varsigma^2)^2 + (\varsigma^3)^2}\right) p, \tag{1}$$

where the terms ς^i depend on photon energy and propagation direction, determined by the SME coefficients, and the two signs correspond to left- and right-handed polarization. The terms ς^0, ς^1, and ς^2 only depend on coefficients of even mass dimension, and ς^3 only depends on odd-mass-dimension coefficients,

$$\varsigma^3 = \sum_{\substack{d \text{ odd} \\ jm}} E^{d-4} \, {}_0Y_{jm}(\hat{\boldsymbol{n}}) k^{(d)}_{(V)jm}. \tag{2}$$

Here, ${}_0Y_{jm}$ are spherical harmonics, $\hat{\boldsymbol{n}}$ is a unit vector pointing in the opposite direction of the photon propagation (i.e., toward the source), and $k^{(d)}_{(V)jm}$ are the SME coefficients. Not all values of j, m are allowed and at

$d = 5$ there are 16 coefficients in total. However, in this chapter, we will only consider the $d = 5$ isotropic term,

$$\varsigma^3_{(5)\text{iso}} = E\,{}_0Y_{00}k^{(5)}_{(V)00}. \tag{3}$$

An extension of the analysis presented here to anisotropic models and other mass dimensions will be the subject of a forthcoming paper.

As a photon propagates, its Stokes vector, $\boldsymbol{s} = (Q, U, V)^T$, will rotate around an axis $\boldsymbol{\varsigma} = (\varsigma^1, \varsigma^2, \varsigma^3)^T$ determined by the terms in the dispersion relation:

$$\frac{d\boldsymbol{s}}{dt} = 2E\boldsymbol{\varsigma} \times \boldsymbol{s}. \tag{4}$$

Because ς^1 and ς^2 vanish for any odd d, the birefringence axis $\boldsymbol{\varsigma}$ points in the V direction, which means that circular polarization V will be unaffected while the linear polarization direction ψ rotates continuously.

Assuming two photons of energies E_1 and E_2 emitted by a source at redshift z have the same polarization angle at the source, the observed polarization angles at Earth will differ by

$$\Delta\psi = (E_1^2 - E_2^2)\,{}_0Y_{00}k^{(5)}_{(V)00} \int_0^z \frac{1 + z'}{H_{z'}} dz' \tag{5}$$

in the $d = 5$ isotropic model, where $H_{z'}$ is the Hubble expansion rate at redshift z'. It is clear from this equation that constraints on $k^{(5)}_{(V)00}$ can be derived by measuring $\Delta\psi$ over some energy range. However, in this paper, we take an alternative approach that will allow us to make use of observations that do not allow energy-resolved polarimetry.

Following the approach developed in Refs. 7 and 9, we continue to assume that the polarization angle at the source is independent of energy. The Stokes parameters of a photon of energy $E_1 \leq E \leq E_2$ within the band pass of IXPE of $E_1 = 2\,\text{keV}$ and $E_2 = 8\,\text{keV}$ at the observer are

$$q(E) = \cos\left(2\zeta^{(5)}_{k00}(E^2 - E_1^2)\right), \tag{6}$$

$$u(E) = \sin\left(2\zeta^{(5)}_{k00}(E^2 - E_1^2)\right), \tag{7}$$

where we choose our coordinate system such that $Q(E_1) = U(E_1) = 0$ and define

$$\zeta^{(5)}_{k00} = {}_0Y_{00}k^{(5)}_{(V)00} \int_0^z \frac{1 + z'}{H_{z'}} dz'. \tag{8}$$

If $\Pi(E)$ is the polarization at the source, the spectrally integrated Stokes parameters of an IXPE observation are

$$Q = \int_{E_1}^{E_2} \Pi(E) \cos\left(2\zeta^{(5)}_{k00}(E^2 - E_1^2)\right) \mu(E)F(E)\, dE, \tag{9}$$

$$U = \int_{E_1}^{E_2} \Pi(E) \sin\left(2\zeta^{(5)}_{k00}(E^2 - E_1^2)\right) \mu(E)F(E)\, dE, \tag{10}$$

where $F(E)$ is the observed flux at energy E, and $\mu(E)$ is the modulation factor of IXPE, which describes the polarization sensitivity of the instrument. The modulation factor is defined as the polarization measured for a fully polarized incident beam. It is easy to see that in general $\zeta^{(5)}_{k00} \neq 0$, corresponding to $k^{(5)}_{(V)00} \neq 0$, will result in a reduced observed polarization fraction $P = \sqrt{Q^2 + U^2}$ compared to the case of no Lorentz invariance violation.

Because $\Pi(E)$ at the source is unknown, we instead assume $\Pi(E) \equiv 1$ to calculate Q_{max} and U_{max}, which lead to the largest observable polarization

$$P_{\text{max}} = \sqrt{Q^2_{\text{max}} + U^2_{\text{max}}} \tag{11}$$

for a given value of $k^{(5)}_{(V)00}$. Any lower polarization at the source will result in a lower observed polarization. Figure 1 shows the maximum observable polarization as a function of $\zeta^{(5)}_{k00}$ for a sample of three IXPE polarization observations of AGNs. We then calculate the likelihood $\mathcal{P}(P < P_{\text{max}}|\sigma_P)$

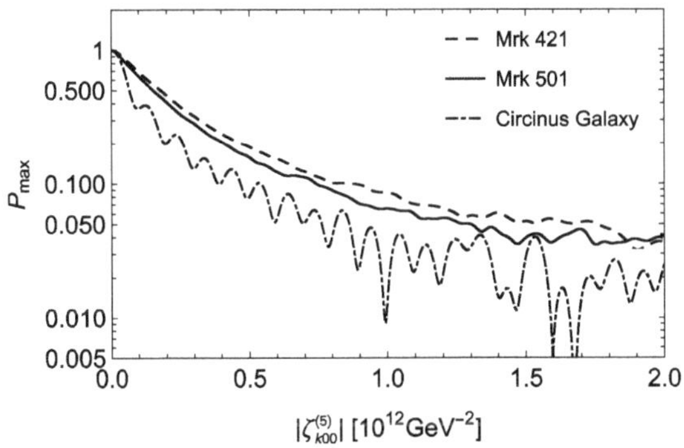

Fig. 1. Theoretically possible maximum observable polarization P_{max} according to Eq. (11) as a function of the Lorentz-violating parameter $\zeta^{(5)}_{k00}$ for three AGN polarization measurements with IXPE. The curves are determined primarily by the measured source spectra $F(E)$.

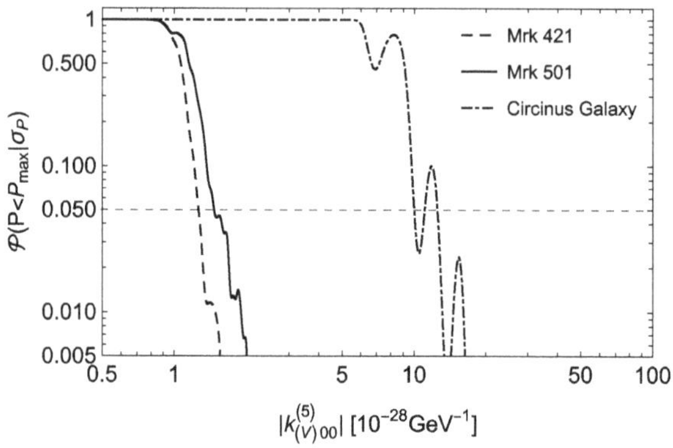

Fig. 2. By comparing IXPE polarization measurements (Mrk 421: $(15 \pm 2)\%$,[11] Mrk 501: $(10 \pm 2)\%$,[12] Cir Galaxy: $(20 \pm 3.8)\%$[13]) to the maximum observable polarization we calculate the likelihood that the true polarization at Earth is less than the maximum possible polarization, $P_{\max}$, shown in Fig. 1. The limit on the SME coefficient $k^{(5)}_{(V)00}$ is then the largest value where $\mathcal{P}(P < P_{\max}|\sigma_P) \geq 0.05$.

that the true polarization value at the observer P is less than $P_{\max}$ given the measured polarization and its uncertainty σ_P.

The resulting likelihood $\mathcal{P}(P < P_{\max}|\sigma_P)$ for the three AGN observations from Fig. 1 is shown in Fig. 2. An upper limit on $k^{(5)}_{(V)00}$ at the 95 % confidence level is then derived by finding the largest value where $\mathcal{P} \geq 0.05$. We find values $k^{(5)}_{(V)00} \lesssim \mathcal{O}(10^{-27}\,\text{GeV}^{-1})$, which is about three orders of magnitude smaller than the limits we previously found for an anisotropic model by combining optical polarimetry of 1278 unique astrophysical objects.

Given the E^2 dependence of $\Delta\psi$ and the fact that the IXPE energy range is a factor 10^3 higher than optical frequencies, one might expect a 10^6 improvement of the result. However, our previous results have the advantage that many more optical polarization measurements are available than X-ray polarization measurements, many of which are at higher redshift, and statistically more significant.

3. Results

Table 1 lists all IXPE polarization detections from 11 different AGNs and Fig. 3 shows their distribution in the sky. It is noteworthy that all are BL Lac-type blazars or Seyfert 1/2 galaxies. For low or intermediate-energy peaked blazars, only upper limits on the polarization fraction have been

Table 1. IXPE polarization detections from 11 active galactic nuclei and resulting constraints on the isotropic SME coefficient $k^{(5)}_{(V)00}$. All redshifts were obtained from the SIMBAD astronomical database at https://simbad.u-strasbg.fr.

Source	Redshift z	Polarization (%)	Ref.	$\left\|k^{(5)}_{(V)00}\right\|_{\max}$ ($10^{-29}\,\mathrm{GeV}^{-1}$)
Mrk 421	0.030	15 ± 2	[11]	12.5
Mrk 501	0.033	10 ± 2	[12]	13.7
Cir Galaxy	0.00145	20 ± 3.8	[13]	120
NGC 1068	0.00348	12.4 ± 3.6	[14]	127
NGC 4151	0.00315	4.9 ± 1.1	[15]	73.9
H1426+428	0.129	20.3 ± 4.1	[16]	1.87
PKS 2155–304	0.116	30.7 ± 2.0	[17]	1.44
IC 4329A	0.0160	3.3 ± 1.1	[18]	83.9
1ES 1959+650	0.047	8.0 ± 2.3	[19]	11.7
1ES 0229+200	0.1397	17.9 ± 2.8	[20]	1.83
PG 1553+113	0.360	10 ± 2	[21]	1.56

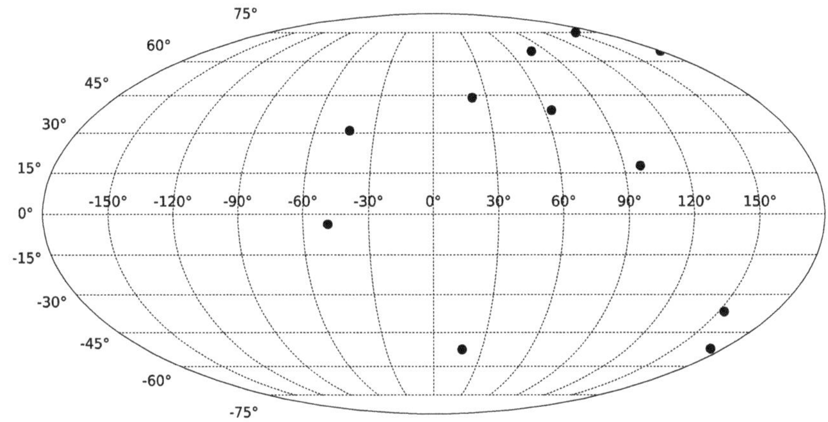

Fig. 3. Skymap in galactic coordinates showing the locations of the 11 AGN from which IXPE has detected non-zero polarization to date.

found. We applied the methods described in the previous section to each of the sources and the resulting upper limits on $k^{(5)}_{(V)00}$ are also given in the table. The strength of the individual constraints is determined primarily by the source redshift z and, to a lesser degree, by the significance of the polarization detection.

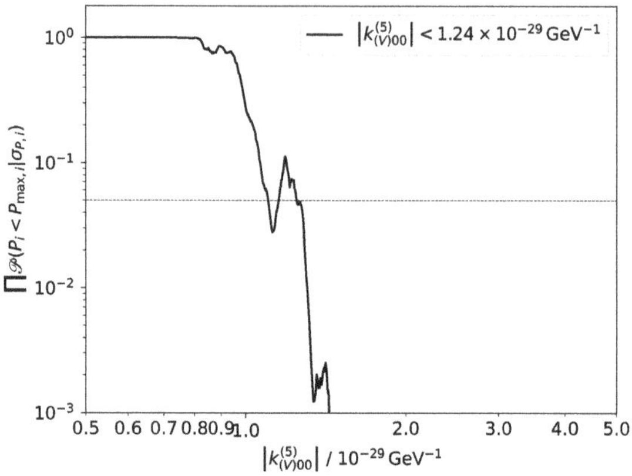

Fig. 4. Product likelihood $\prod_i \mathcal{P}(P_i < P_{\max,i}|\sigma_{P,i})$ that all IXPE measurements listed in Table 1 are compatible with a given value of $k^{(5)}_{(V)00}$. At the 95 % confidence level, this leads to the upper limit $\left|k^{(5)}_{(V)00}\right| < 1.24 \times 10^{-29}\,\text{GeV}^{-1}$.

The individual results can be combined by calculating the product likelihood

$$\prod_i \mathcal{P}(P_i < P_{\max,i}|\sigma_{P,i})$$

which is shown in Fig. 4. We find the upper limit

$$\left|k^{(5)}_{(V)00}\right| < 1.24 \times 10^{-29}\,\text{GeV}^{-1},$$

which is dominated by the strongest individual constraint, as expected. Although no significant improvement from using multiple sources is found in this isotropic analysis, this approach will make a significant difference in an anisotropic analysis, where all $k^{(d)}_{(V)jm}$ for any given d are constrained simultaneously.

4. Summary and Outlook

Using IXPE X-ray polarization measurements of AGNs in the 2–8 keV energy range, we derived a new constraint on the isotropic $d = 5$ SME coefficient in the photon sector. IXPE is the first dedicated X-ray polarimetry mission in more than 40 years and the first such mission that has measured X-ray polarization from an extragalactic source. While claims of gamma-ray polarization of GRBs have previously been made, those measurements suffer from large systematic uncertainties because the instruments were not calibrated for polarization measurements. These detections are also in some tension with the non-detection of gamma-ray polarization from GRBs with the dedicated GRB polarimeter POLAR.[22] While constraints derived from these measurements are in principle stronger than those presented here,[5,6,23] they are affected by these systematic uncertainties. The constraints presented here do not suffer from the same issue because IXPE has been carefully calibrated for polarimetry prior to launch.

In a future paper, we will use the same data in combination with optical polarization measurements to constrain all SME photon-sector coefficients of $d = 5$ and possibly higher mass dimension using the techniques described in Ref. 9.

While IXPE polarization detections from additional extragalactic sources will continue to tighten possible constraints, another significant improvement in sensitivity is to be expected from polarimetry at higher energy enabled by proposed gamma-ray burst polarimetry missions or the upcoming Compton Spectrometer and Image (COSI) mission.

Acknowledgments

This work is funded in part by NASA grants 80NSSC24K0205, 80NSSC24K0636, and 80NSSC24K1762.

References

1. V.A. Kostelecký and N. Russell, 2025 edition, *Data Tables for Lorentz and CPT Violation*, arXiv:0801.0287v18.
2. R.C. Myers and M. Pospelov, *Phys. Rev. Lett.* **90**, 211601 (2003); T.G. Rizzo, *J. High Energy Phys.* **2005**, 036 (2005); V.A. Kostelecký and S. Samuel, *Phys. Rev. D* **39**, 683 (1989); C.P. Burgess *et al.*, *J. High Energy Phys.* **2002**, 043 (2002); R. Gambini and J. Pullin, *Phys. Rev. D* **59**, 124021 (1999); M. Pospelov and Y. Shang, *Phys. Rev. D* **85**, 105001 (2012); M. Li, Y.F. Cai, X. Wang and X. Zhang, *Phys. Lett. B* **680**, 118 (2009).
3. V. Vasileiou *et al.*, *Phys. Rev. D* **87**, 122001 (2013); F. Kislat and H. Krawczynski, *Phys. Rev. D* **92**, 045016 (2015); J.J. Wei and X.F. Wu, *Front. Phys.* **16**, 44300 (2021).
4. P. Kaaret, Nature **427**, 287 (2004);
5. V.A. Kostelecký and M. Mewes, *Phys. Rev. Lett.* **110**, 201601 (2013).
6. V.A. Kostelecký and M. Mewes, *Phys. Rev. D* **80**, 015020 (2009).
7. F. Kislat and H. Krawczynski, *Phys. Rev. D* **95**, 083013 (2017); F. Kislat, Symmetry **10**, 596 (2018); R. Gerasimov, P. Bhoj, and F. Kislat, *Symmetry* **13**, 880 (2021).

8. D. Colladay and V.A. Kostelecký, *Phys. Rev. D* **55**, 6760 (1997); D. Colladay and V.A. Kostelecký, *Phys. Rev. D* **58**, 116002 (1998); V.A. Kostelecký and M. Mewes, *Phys. Rev. D* **66**, 056005 (2002); V.A. Kostelecký, *Phys. Rev. D* **69**, 105009 (2004).
9. A.S. Friedman *et al.*, *Phys. Rev. D* **102**, 043008 (2020).
10. M. Weisskopf *et al.*, *J. Astron. Telescopes Instrum. Syst.* **8**, 026002 (2022).
11. L. Di Gesu *et al.*, *Astrophys. J. Lett.* **938**, L7 (2022).
12. I. Liodakis *et al.*, *Nature* **611**, 677 (2022).
13. F. Ursini *et al.*, *Mon. Not. Royal Astron. Soc.* **519**, 50 (2023).
14. F. Marin *et al.*, *Astron. Astrophys.* **689**, A238 (2024).
15. V.E. Gianolli *et al.*, *Mon. Not. Royal Astron. Soc.* **523**, 4468 (2023); V.E. Gianolli *et al.*, *Astron. Astrophys.* **691**, A29 (2024).
16. A. Banerjee *et al.*, arXiv:2504.12410.
17. P.M. Kouch *et al.*, *Astron. Astrophys.* **689**, A119 (2024).
18. A. Ingram *et al.*, *Mon. Not. Royal Astron. Soc.* **525**, 5437 (2023).
19. M. Errando *et al.*, *Astrophys. J.* **963**, 5 (2024).
20. S.R. Ehlert *et al.*, *Astrophys. J.* **959**, 61 (2023).
21. R. Middei *et al.*, *Astrophys. J. Lett.* **942**, L10 (2023).
22. M. Kole *et al.*, *Astron. Astrophys.* **644**, A124 (2020).
23. K. Toma *et al.*, *Phys. Rev. Lett.* **109**, 241104 (2012); P. Laurent *et al.*, *Phys. Rev. D* **83**, 121301(R) (2011); F.W. Stecker, *Astropart. Phys.* **35**, 95 (2011).

Chapter 6

Nonminimal Lorentz Violation in Atomic and Molecular Spectroscopy Experiments

Arnaldo J. Vargas

Laboratory of Theoretical Physics, Department of Physics, University of Puerto Rico, Río Piedras 00936, Puerto Rico

This presentation discusses potential signals of Lorentz violation that could be observed in atomic and molecular spectroscopy experiments. It provides a general overview of the nonrelativistic effective SME coefficients that have been constrained, as well as the prospects for placing first-time bounds on those that remain unconstrained. Additionally, it highlights the importance of considering transitions involving atomic or molecular states with high angular momentum in tests of Lorentz symmetry.

1. Introduction

The possibility that tiny deviations from Lorentz symmetry could serve as low-energy signals of physics beyond the Standard Model and General Relativity[1] led to the development of the Standard-Model Extension (SME),[2] a comprehensive framework for systematically testing Lorentz symmetry. The SME assigns a unique coefficient for Lorentz violation, known as an SME coefficient, to each Lorentz-violating operator. The goal of the SME is to search for evidence of a nonzero SME coefficient, which would indicate a violation of Lorentz symmetry.

The SME also provides a framework for systematically testing CPT symmetry, since CPT violation implies Lorentz violation in a realistic effective quantum field theory.[3] Approximately, half of the Lorentz-violating operators in the SME also violate CPT symmetry. The coefficients associated with these operators are referred to as coefficients for CPT violation, and any evidence of a nonzero value for such a coefficient would suggest CPT violation.

Experiments that probe these symmetries are typically sensitive to a limited subset of SME coefficients. Therefore, a wide range of experiments is necessary for the systematic testing of Lorentz and CPT symmetry. Using the SME framework, one can quantify the sensitivity of an experiment to specific SME coefficients, which allows for comparing the competitiveness and complementarity of different tests of these symmetries.

Early versions of the SME focused on Lorentz-violating operators of mass dimension $d \leq 4$, referred to as minimal operators, to avoid introducing nonrenormalizable terms into the framework.[2] However, this restriction is not essential, since the SME is based on the premise that any underlying Lorentz-violating theory would manifest as a low-energy effective field theory. Such a theory does not need to be renormalizable if it is only valid within a limited energy range. Consequently, there is no reason to exclude nonminimal operators, defined as those with $d \geq 5$, from the SME. Subsequent extensions of the SME have included these nonminimal operators.[4–8] In some ways, the nonminimal operators are more interesting than the minimal ones, as their effects are expected to be suppressed by the ratio of the experimental energy scale to the high-energy scale at which the effective theory breaks down, providing a plausible explanation for the tiny effects, currently unobserved, of Lorentz and CPT violation.

The following presents an overview of models for testing Lorentz and CPT symmetry in atomic and molecular spectroscopy experiments that incorporate nonminimal effects, with particular emphasis on those developed in Refs. 9–15. Owing to the focused scope of this discussion, we do not consider high-precision spectroscopy experiments that test models involving only minimal SME coefficients and have not been used to impose bounds on nonminimal ones.[16]

In many cases, the contributions from SME coefficients to the observables of a system appear in specific linear combinations. For convenience, these linear combinations are defined as effective SME coefficients. In the model discussed here, the effective coefficients are the nonrelativistic (NR) coefficients.[6]

The structure of this presentation is as follows. We provide a brief description of the NR coefficients and their relevant properties. We then discuss potential signals of Lorentz violation that could be observed in atomic and molecular experiments. Also, we review the NR coefficients that have already been constrained and highlight potential experiments that could impose bounds on the remaining unconstrained NR coefficients.

2. Main

The dominant Lorentz-violating effects on the atomic spectrum can be obtained by considering only the leading Lorentz-violating perturbation to the free propagation of the fermions within the atom.[10] The perturbation of interest to us is the one obtained by adding a single-particle perturbation derived in Ref. 6, containing both minimal and nonminimal operators, for each fermion in the atom. The perturbation depends on the free-particle energy $\sqrt{|\boldsymbol{p}|^2+m_w^2}$, which can be expanded in powers of $|\boldsymbol{p}|/m_w$ if it is smaller than one. The resulting perturbation, obtained by applying this expansion to the free-particle energy, is called the nonrelativistic (NR) perturbation, denoted by $\delta h_w^{\rm NR}$. Here, w denotes the fermion type and takes the values e for electron, p for proton, and μ for muon.

As demonstrated in Ref. 10, some terms in the single-particle perturbation do not contribute to the energy of atomic or molecular levels. After excluding these terms, the Lorentz-violating single-particle perturbation takes the form

$$\begin{aligned}\delta h_w^{\rm NR} &= -\sum_{kjm} |\boldsymbol{p}|^k {}_0Y_{jm}(\hat{\boldsymbol{p}}) \left(\mathcal{V}_{w\,kjm}^{\rm NR} + \boldsymbol{\sigma}\cdot\hat{\boldsymbol{\epsilon}}_r \mathcal{T}_{w\,kjm}^{{\rm NR}(0B)}\right) \\ &\quad + \sum_{kjm} |\boldsymbol{p}|^k \boldsymbol{\sigma}\cdot\left({}_{+1}Y_{jm}(\hat{\boldsymbol{p}})\hat{\boldsymbol{\epsilon}}_- - {}_{-1}Y_{jm}(\hat{\boldsymbol{p}})\hat{\boldsymbol{\epsilon}}_+\right)\mathcal{T}_{w\,kjm}^{{\rm NR}(1B)}, \qquad (1)\end{aligned}$$

where k takes the values 0, 2, and 4 with $0 \le j \le 5$ and $-j \le m \le j$.

The use of the nonrelativistic expansion of the free-particle energy leads to linear combinations of SME coefficients of different mass dimensions in the perturbation, each accompanied by appropriate powers of the fermion mass. These linear combinations of coefficients are called the nonrelativistic (NR) coefficients and are distinguished by the superscript NR. The relation between the NR coefficients and the standard SME coefficients for Lorentz and CPT violation is provided in Eqs. (111) and (112) of Ref. 6. In general, all NR coefficients receive contributions from nonminimal SME coefficients, while some also receive contributions from minimal ones.

The NR coefficients appearing in $\delta h_w^{\rm NR}$, namely, $\mathcal{V}_{w\,kjm}^{\rm NR}$, $\mathcal{T}_{w\,kjm}^{\rm NR(0B)}$ and $\mathcal{T}_{w\,kjm}^{\rm NR(1B)}$, are linear combinations of SME coefficients associated with operators with a definite CPT sign. These combinations are

$$\begin{aligned}\mathcal{V}_{w\,kjm}^{\rm NR} &= c_{w\,kjm}^{\rm NR} - a_{w\,kjm}^{\rm NR},\\ \mathcal{T}_w{}^{\rm NR(qP)}kjm &= g_w{}^{\rm NR(qP)}kjm - H_{w\,kjm}^{\rm NR(qP)},\end{aligned} \tag{2}$$

where the a- and g-type coefficients correspond to CPT-odd operators, while the c- and H-type coefficients are associated with CPT-even ones. In contrast, nonrelativistic antimatter experiments are sensitive to a different linear combination. To distinguish between these two scenarios, the index $\overline{w}$ is used to denote the antiparticle case, in contrast to w, which refers to the particle one. The combinations relevant to antimatter experiments are given by

$$\begin{aligned}\mathcal{V}_{\overline{w}kjm}^{\rm NR} &= c_{w\,kjm}^{\rm NR} + a_{w\,kjm}^{\rm NR},\\ \mathcal{T}_{\overline{w}}{}^{\rm NR(qP)}kjm &= -g_w{}^{\rm NR(qP)}kjm - H_{w\,kjm}^{\rm NR(qP)}.\end{aligned} \tag{3}$$

The difference between Eqs. (2) and (3) lies in the sign of the CPT-odd terms. It is important to note that coefficients with definite CPT sign are always defined relative to the particle type, not the antiparticle one, which is consistent with standard conventions in the literature. Finally, the indices j and m of the NR coefficients correspond to those of the spin-weighted spherical harmonics ${}_sY_{jm}(\hat{\boldsymbol{p}})$ appearing in Eq. (1), while the index k indicates the power of the fermion's momentum magnitude $|\boldsymbol{p}|$.

Not all combinations of the k and j indices are allowed in the perturbation (1). For the $\mathcal{V}$-type coefficients, only even values of j are permitted, and k must satisfy $k \geq j$. For the $\mathcal{T}$-type coefficients, j is odd and must satisfy $k+1 \geq j$. Overall, the model based on the perturbation (1) contains 178 nonrelativistic (NR) coefficients accessible in atomic spectroscopy experiments, per particle type. For example, hydrogen spectroscopy experiments are sensitive to 178 electron coefficients and an equal number of proton coefficients.

It is important to note that not all 178 coefficients contribute to the Lorentz-violating energy shift for a given atomic level, as discussed in Ref. 10. For instance, in two-fermion atoms such as hydrogen, antihydrogen, and muonium, only coefficients with index j satisfying $j \leq 2J$ and $j \leq 2F$ can contribute to a level characterized by quantum number J (the total angular momentum of the lighter fermion) and F (the total atomic angular momentum).[10] For instance, transitions between nS states in two-fermion atoms with quantum numbers $J = 1/2$ and $F \leq 1$ are sensitive to only 40 coefficients per particle type—a small fraction of the total number of coefficients included in the model.

If rotational symmetry is broken, the atomic spectrum may depend on the orientation of the total angular momentum $\vec{F}$ of the atom. This could lead to unexpected multiple resonance peaks or broadening of the line shape.[9,10,14] However, the most common experimental signal of rotational symmetry violation in Earth-based experiments is a sidereal variation of the resonant frequency. The basic idea is that the atomic spectrum changes when the atom, specifically its total angular momentum $\vec{F}$, is rotated relative to a fixed inertial reference frame. This rotation can be achieved by applying an external magnetic field that is adiabatically rotated with respect to a fixed frame. A signal of rotational symmetry breaking would then manifest as a time variation in the resonance frequencies of the atom, modulated at the rotation frequency of the magnetic field. For experiments conducted on the surface of the Earth, it is sufficient to fix the magnetic field in the laboratory frame, since the Earth's rotation naturally produces an adiabatic rotation relative to a fixed inertial frame. In this case, the modulation frequency corresponds to the sidereal frequency.

The SME coefficients transform under observer transformations; therefore, their values depend on the reference frame used to analyze the experiment. To enable comparison across multiple experiments, it is standard practice to report constraints on the coefficients in a common frame, typically the Sun-centered frame.[17,18] The indices j and m of the NR coefficients in the Sun-centered frame that contribute to the Lorentz-violating shift in the resonant frequency provide insight into the type of sidereal variation expected in an experiment. Suppose the time variation of the transition frequency is decomposed into harmonics of the sidereal frequency, $\omega_\oplus \simeq 2\pi/(1436\,\text{min})$. An NR coefficient in the Sun-centered frame with index m can contribute only to the amplitude of the $|m|$th harmonic of $\omega_\oplus$.[10] Since $|m| \leq j$, a coefficient with index j cannot contribute to harmonics higher than the jth harmonic of the sidereal frequency.

The NR coefficients with $j \leq 2$ contain the dominant contributions from the minimal SME coefficients, specifically the $\mathcal{T}$-type with $k = 0$ for $j = 1$ and the $\mathcal{V}$-type with $k = 2$ for $j = 0$ and $j = 2$. For this reason, many experiments motivated by the minimal SME focused on these types of coefficients, and the constraints obtained from those experiments have since been reinterpreted as bounds on the NR coefficients with $j = 1$ and $j = 2$.

In particular, studies of sidereal variation in resonance frequencies within the ground state of muonium[19] and hydrogen,[20] as well as experiments using a Xe-He co-magnetometer,[21] have been used to place bounds on the $j = 1$ $\mathcal{T}$-type NR coefficients with $|m| = 1$ in the electron, proton, muon, and neutron sectors of the SME.[9–11] A sidereal variation study using a Ne-Rb-K co-magnetometer,[22] which investigated variations at the first two harmonics of the sidereal frequency, was used to constrain the $j = 2$ $\mathcal{V}$-type NR coefficients for the neutron, with $|m| = 1$ and $|m| = 2$.[11]

NR coefficients with index $j \geq 3$ remain unconstrained,[18] as all transitions considered thus far have involved states with total angular momentum less than or equal to $F = 1$. Consequently, Lorentz and CPT tests involving transitions to states with higher angular momentum could offer sensitivity to these unconstrained NR coefficients. For example, the proposed measurement of the $2S$–$2P_{3/2}$ transition in muonium could provide the first bounds on $j = 3$ muon NR coefficients[14] and sidereal variation studies of the rovibrational transitions[23] within the ground state of H_2^+ could impose the first bounds on the proton NR coefficients with $j = 4$.[15]

Another crucial feature of the perturbation (1) is its dependence on the magnitude of the momentum, which allows for some systems to exhibit significantly greater sensitivity to certain coefficients, particularly those with $k = 4$, due to the higher momentum of their constituents. For example, the proton in deuterium has a higher momentum than in hydrogen because of its motion within the deuteron. Consequently, deuterium can be significantly more sensitive to proton coefficients than hydrogen, even though hydrogen typically offers better sensitivity to Lorentz-violating frequency shifts.[10,12] A sidereal variation study of the Zeeman-hyperfine transitions within the ground state of deuterium, currently in its final stages of analysis, is expected to improve bounds on the proton NR coefficients with $j = 1$ for $k = 2$ and $k = 4$, and to provide the first constraints on the proton NR coefficients with $j = 2$.[12] Another system benefiting from this momentum enhancement is muonic hydrogen, in which the negative muon has greater momentum than the positive muon in muonium. Therefore, although the proposed improved measurement of the hyperfine splitting in the ground state of muonium[24] is the best candidate for improving bounds on the $\mathcal{T}$-type $k = 0$ muon coefficients, the best prospects for tightening bounds on the $k = 2$ and $k = 4$ muon coefficients lie in sidereal variation studies using the data collected in muonic hydrogen experiments,[9] such as the one currently underway.[25] Finally, rovibrational transitions within the ground state of HD^+ have also been measured experimentally.[26] Owing to the higher momentum of the proton within the deuteron, these transitions may offer enhanced sensitivity to proton coefficients compared to those in H_2^+.[15]

Nonrelativistic (NR) coefficients in the Sun-centered frame with $m = 0$ do not induce sidereal variations in the resonance frequency.[9–11] However, these coefficients can still lead to constant frequency shifts, which can be constrained through alternative methods. One such method exploits the dependence of certain $m = 0$ coefficients on the relative orientation between the magnetic field and Earth's rotation axis.[10] An experiment studying transitions within the ground state of hydrogen for different magnetic field orientations was able to place limits on the $\mathcal{T}$-type proton and electron coefficients with $j = 1, m = 0$.[27] Another strategy for probing $m = 0$ coefficients involves tests of CPT symmetry.[10] For example, comparisons between the $1S$–$2S$ transitions in hydrogen and antihydrogen[28] have yielded bounds on the a-type coefficients with $j = 0$ for both electrons and protons.[11] Additionally, coefficients with $m = 0$ can be constrained by comparing precise experimental results with predictions from the Standard Model. This approach has led to bounds on a-type and c-type coefficients in the electron, proton, and muon sectors. Specific examples include the $1S$–$2S$ transition in positronium,[10] the $1S$–$2S$ transition and classical Lamb shift in muonium,[9,29] and the $1S$–$2P$ transition in antihydrogen.[13,30] As with sidereal variation studies, it is desirable to investigate transitions involving states with higher angular momentum quantum numbers. In this context, the proposed possibility of performing molecular antihydrogen spectroscopy[23,31] is particularly exciting, as it offers sensitivity to NR coefficients with $j \geq 3$ and $m = 0$, which remain inaccessible in the transitions studied thus far in antihydrogen.[10,15]

3. Final Remarks

Among the NR coefficients considered in the models discussed here, only about 16% have been constrained in the muon sector and roughly 25% in the electron, proton, and neutron sectors. This leaves substantial room for exploring possible violations of CPT and Lorentz symmetry in atomic and molecular experiments. A key reason for the large number of unconstrained NR coefficients is the scarcity of Lorentz and CPT tests involving transitions with high angular momentum states. While it remains important to continue tightening existing bounds, many of which can still be significantly improved, it is equally important to design and carry out experiments that specifically probe the currently unconstrained coefficients.

References

1. V.A. Kostelecký and S. Samuel, *Phys. Rev. D* **39**, 683 (1989); V.A. Kostelecký and R. Potting, *Nucl. Phys. B* **359**, 545 (1991); *Phys. Rev. D* **51**, 3923 (1995).
2. D. Colladay and V.A. Kostelecký, *Phys. Rev. D* **55**, 6760 (1997); *Phys. Rev. D* **58**, 116002 (1998).
3. O.W. Greenberg, *Phys. Rev. Lett.* **89**, 231602 (2002).
4. V.A. Kostelecký and M. Mewes, *Phys. Rev. D* **80**, 015020 (2009).
5. V.A. Kostelecký and M. Mewes, *Phys. Rev. D* **85**, 096005 (2012); J.S. Díaz, V.A. Kostelecký, and M. Mewes, *Phys. Rev. D* **89**, 043005 (2014); V.A. Kostelecký and Z. Li, *Phys. Rev. D* **99**, 056016 (2019).
6. V.A. Kostelecký and M. Mewes, *Phys. Rev. D* **88**, 096006 (2013).
7. V.A. Kostelecký and M. Mewes, *Phys. Lett. B* **757**, 510 (2016); Q.G. Bailey, V.A. Kostelecký, and R. Xu, *Phys. Rev. D* **91**, 022006 (2015); V.A. Kostelecký and J.D. Tasson, *Phys. Lett. B* **749**, 551 (2015).
8. V.A. Kostelecký and Z. Li, *Phys. Rev. D* **103**, 024059 (2021).
9. A.H. Gomes, V.A. Kostelecký, and A.J. Vargas, *Phys. Rev. D* **90**, 076009 (2014).
10. V.A. Kostelecký and A.J. Vargas, *Phys. Rev. D* **92**, 056002 (2015).
11. V.A. Kostelecký and A.J. Vargas, *Phys. Rev. D* **98**, 036003 (2018).
12. A.J. Vargas, *Phys. Rev. D* **109**, 055001 (2024).
13. A.J. Vargas, *Phys. Rev. D* **111**, 015032 (2025).
14. P. Blumer *et al.*, *Eur. Phys. J. D* **79**, 24 (2025).
15. A.J. Vargas, arXiv:2503.06306 [hep-ph].
16. M.A. Hohensee *et al.*, *Phys. Rev. Lett.* **111**, 050401 (2013); T. Pruttivarasin *et al.*, *Nature* **517**, 592 (2015); C. Sanner *et al.*, *Nature* **567**, 204 (2019); Eli Megidish *et al.*, *Phys. Rev. Lett.* **122**, 123605 (2019); L.S. Dreissen *et al.*, *Nat. Commun.* **13**, 7314 (2022).
17. V.A. Kostelecký and M. Mewes, *Phys. Rev. D* **66**, 056005 (2002).
18. V.A. Kostelecký and N. Russell, *Rev. Mod. Phys.* **83**, 11 (2011); *Data Tables for Lorentz and CPT Violation,* 2025 edition, arXiv:0801.0287v18.
19. V.W. Hughes *et al.*, *Phys. Rev. Lett.* **87**, 111804 (2001).
20. D.F. Phillips *et al.*, *Phys. Rev. D* **63**, 111101(R) (2001); M.A. Humphrey *et al.*, Phys. Rev. A **68**, 063807 (2003).
21. F. Allmendinger *et al.*, *Phys. Rev. Lett.* **112**, 110801 (2014).
22. M. Smiciklas *et al.*, *Phys. Rev. Lett.* **107**, 171604 (2011).
23. M.R. Schenkel, S. Alighanbari, and S. Schiller, *Nat. Phys.* **20**, 383 (2024).
24. P. Strasser *et al.*, EPJ Web of Conferences **198**, 00003 (2019).
25. C. Pizzolotto *et al.*, Eur. Phys. J. A **56**, 185 (2020); P. Amaro *et al.*, *SciPost Phys.* **13**, 020 (2022).

26. S. Alighanbari *et al.*, Nature **581**, 152 (2020); S. Patra *et al.*, *Science* **369**, 1238 (2020).
27. L. Nowak *et al.*, *Phys. Lett. B* **854**, 139012 (2024).
28. M. Ahmadi *et al.*, *Nature* **557**, 71 (2018); Nature **541**, 506 (2017).
29. B. Ohayon *et al.*, *Phys. Rev. Lett.* **128**, 011802 (2022).
30. M. Ahmadi *et al.*, *Nature* **578**, 375 (2020).
31. H. Dehmelt, *Phys. Scr.* **T59**, 423 (1995); E.G. Myers, *Phys. Rev. A* **98**, 010101 (2018); M.C. Zammit *et al.*, *Phys. Rev. A* **100**, 042709 (2019).

Chapter 7

Neutral-Meson to Light-Quark Connection of the CPT-Violating Coefficients in the Nonminimal Standard-Model Extension

A. Roberts

Indiana University Center for Spacetime Symmetries,
Bloomington, IN 47405, USA

The theoretical description of CPT violation in neutral-meson systems to arbitrary mass dimension by scalar effective field theory has been known for some time. This presentation describes the connection between this scalar field theory at the meson level and the more fundamental quark-level CPT breaking in the Standard-Model Extension, relating the relevant CPT-violating coefficients in these two theoretical approaches. For mesons containing light quarks, specifically the neutral kaon of down and strange quarks, this is achieved using Chiral Perturbation Theory with three approximately massless light quarks. Details are examined for mass dimension five in the phenomenology of neutral-kaon experiments.

1. Introduction

Effective field theories (EFTs), such as the Standard-Model Extension (SME), can be implemented when the full underlying theory is unknown or experimentally inaccessible. They can be suitable for producing verifiable phenomenology appropriate for a domain of investigation. These domains can have different degrees of freedom from the underlying theory and are valid at specified scales. However, to calculate needed observables the effective Lagrangian of any such EFT must be constructed such that it is the most general description consistent with the symmetries of the more fundamental theory.

The SME itself is an EFT that respects the gauge symmetries of the Standard Model (SM) but extends it with Lorentz- and CPT-violating terms formed from Lorentz-violating (LV) coefficients contracted with SM fields into Lorentz scalars. These coefficients themselves can refer to a different sector depending on the observable values relevant in experimental searches. With the conception of the SME, neutral-meson systems (NMS), in which discrete symmetries had been observed to be violated, attained great interest.

CPT-violation (CPTV) tests in the NMS are based on interferometry and rely on a Schrödinger-like quantum-mechanics description. They provide information on the SME coefficient Δa^0 for the two differently flavored constituent quarks contained in the meson.[1,2] The issue of not knowing the details of the bound quarks in the meson already surfaced earlier when the general fermion bilinear formulation was introduced.[3] Not knowing the details at the QCD level of the constituent quarks can be bypassed when the mesons, which are physically accessible at low energies, are described as part of the pseudoscalar nonet in a scalar EFT.[4]

The topic of this talk is to connect the established description of CPTV at the meson level by a scalar EFT to the more fundamental form of CPTV at the quark level within the SME. This can be achieved via Chiral Perturbation Theory (χPT) applied to the three lightest quarks, in which the lightest mesons arise as Nambu–Goldstone (NG) bosons of chiral symmetry breaking.[5]

2. CPTV Observable of Neutral-Meson Oscillations

Although well established in the literature, it is wise to identify the observable used to bound CPTV in NMSs. The quantum-mechanical description in the Weisskopf–Wigner approximation appears as[6]

$$i\partial_t \Psi = \Lambda \Psi, \tag{1}$$

where $\Psi(t)$ is a two-dimensional column vector in the meson-antimeson state space. The corresponding 2×2 effective Hamiltonian Λ governing the time evolution of the system can be parametrized as[7]

$$\Lambda = \frac{1}{2}\Delta\lambda \begin{pmatrix} U + \xi & VW^{-1} \\ VW & U - \xi \end{pmatrix}. \tag{2}$$

In this parametrization, U, V, W, and ξ are dimensionless complex coefficients, with $\Delta\lambda \equiv \lambda_a - \lambda_b$ the eigenvalue difference of the effective Hamiltonian Λ. Indirect CPTV is governed by finite values of ξ, which is the experimental observable of our present interest.

Traditionally, ξ is tied to the diagonal-element difference $\Delta\Lambda = \frac{1}{2}(\Lambda_{11} - \Lambda_{22})$ of the Hamiltonian. In a first-order approximation in the coupling constant to the CPTV background this signifies energy-shift differences in the quark propagators[1,3,7]

$$\xi = \frac{\Delta\Lambda}{\Delta\lambda} \, . \tag{3}$$

Additionally, the modulus w of $W \equiv w \exp(i\omega)$ governs T violation with $w = 1$ if and only if T is preserved. See Ref. 7 for more details.

It is important that this description, as well as the parameter ξ, are given in the system's rest frame. Physically, a nonzero ξ indicates CPT violation. In the SME, this manifests in CPT-odd terms, and in a realistic EFT extension of the SM this implies the violation of Lorentz invariance,[8] i.e., the violation of symmetry under boosts and rotations. Hence, in general ξ is boost- and orientation-dependent.[1]

At different levels of approaches, ξ will relate to different coefficients. Since the bounds on ξ are the experimental information, it is taken as the relevant value for the coefficient specified at the mass dimension chosen, assuming all other mass-dimension contributions to be zero. In this presentation, the focus is on $d = 5$ of the nonminimal SME.

3. Meson- and Quark-Level Lagrangians

To relate meson- and quark-level coefficients one can start with the free scalar-level and quark-level Lagrangians. Details of the scalar EFT were worked out for any mass dimension in Ref. 4. Restriction to $d = 5$ leads to the form

$$\mathcal{L}_{\text{SPT}} \supset -\frac{1}{2}\phi^\dagger (k_a^{(5)})^{\mu\alpha\beta} \partial_\mu i\partial_\alpha i\partial_\beta \phi + \text{h.c.} \tag{4}$$

For the quark level, one can adapt the work of Ref. 9 for mass dimension five fermion Lagrangians replacing the covariant derivative with a partial derivative for freely propagating quarks. This matches the approach taken in the experimental context where first-order CPTV is considered to enter through the free propagation of the flavor-specific quark and antiquark, adding effects to the usual weak-force induced oscillations that may be detected with sensitive interferometric methods.

The free SME quark Lagrangian

$$\mathcal{L}_{\mathrm{qSME}} \supset -\frac{1}{2}\bar{\Psi} a^{(5)\mu\alpha\beta}\gamma_\mu i\partial_\alpha i\partial_\beta \chi + \mathrm{h.c.} \tag{5}$$

is the starting point for $d = 5$. The aim is to relate $(k_a^{(5)})^{\mu\alpha\beta}$ to $a^{(5)\mu\alpha\beta}$. By examining the separate C, P, and T properties of coefficients corresponding in the number of time indices, a match between the two sets can be verified, and all coefficient-component groups evaluated according to the time components contained are CPT odd.

An elegant way to establish a relation between $(k_a^{(5)})^{\mu\alpha\beta}$ and $a^{(5)\mu\alpha\beta}$ is to use the light-quark approximation and χPT, where the mesons become the NG bosons of chiral-symmetry breaking. This method was applied previously to connect hadrons and quarks for the two lightest quarks and the minimal renormalizable $d = 3, 4$ terms of the SME.[10]

4. χPT Application to Approximately Massless Light Quarks

The next step now is to find the Lagrangian in the 3-dimensional quark-flavor space of the lightest quarks. It is expected that this Lagrangian follows electroweak gauge symmetry. To do this, it must be formulated in a right- and left-handed decomposition. This appears in the form

$$\delta\mathcal{L}^{a^{(5)}}_{\mathrm{qSME}} = -\bar{\Psi}_L (a_L)^{\mu\alpha\beta}\gamma_\mu i\partial_{(\alpha} i\partial_{\beta)}\Psi_L - \bar{\Psi}_R (a_R)^{\mu\alpha\beta}\gamma_\mu i\partial_{(\alpha} i\partial_{\beta)}\Psi_R\,, \tag{6}$$

with the left-handed column vectors

$$\Psi_L \equiv (u_L, d_L, s_L)^T, \tag{7}$$

and the right-handed ones

$$\Psi_R \equiv (u_R, d_R, s_R)^T. \tag{8}$$

Equation (6) shows the introduction of coefficient matrices $(a_L)^{\mu\alpha\beta}$ and $(a_R)^{\mu\alpha\beta}$ given in u, d, s flavor space. These matrices carry much of the properties left over from the electroweak symmetry which forbids mixing between fields of different charge and refer back to right-handed weak singlets and left-handed weak doublets. With this in mind they are of the form

$$
\begin{aligned}
(a_L)^{\mu\alpha\beta} &\equiv \begin{pmatrix} (a_V^{(5)})_{11} & 0 & 0 \\ 0 & (a_Q^{(5)})_{11} & (a_Q^{(5)})_{12} \\ 0 & (a_Q^{(5)})^*_{12} & (a_Q^{(5)})_{22} \end{pmatrix}^{\mu\alpha\beta}, \\
(a_R)^{\mu\alpha\beta} &\equiv \begin{pmatrix} (a_U^{(5)})_{11} & 0 & 0 \\ 0 & (a_D^{(5)})_{11} & (a_D^{(5)})_{12} \\ 0 & (a_D^{(5)})^*_{12} & (a_D^{(5)})_{22} \end{pmatrix}^{\mu\alpha\beta}.
\end{aligned}
\tag{9}
$$

The $(a_V^{(5)})_{11}$ and $(a_Q^{(5)})_{11}$ entries are equal when restricted to mass dimension five. Taking the light quarks to have zero mass, QCD produces spontaneous breaking of the $SU(3)_L \times SU(3)_R$ chiral symmetry. The members of the lowest-mass pseudoscalar meson octet then present as composite hermitian field operators combined from the NG boson fields of the symmetry breaking.

An $\mathrm{SU(3)_V}$ symmetry is left over as the symmetry of the NG bosons. Referring for details to Ref. 5, the fields from which to construct the relevant meson Lagrangian can be written down as

$$
U = \exp\left(\frac{i\phi}{F_0}\right), \tag{10}
$$

where ϕ is expanded in the Gell-Mann matrices of $SU(3)$, and the NG fields; the details are left to an upcoming publication. Here, it is noted that F_0 stands for the pion decay constant with $4\pi F_\pi \approx 1.17\,\mathrm{GeV}$.

The mass of m_ρ, the lowest-mass meson that is not an NG boson, and F_π set an energy scale for the EFT as $\approx 1\,\mathrm{GeV}$. It is important to note that a nonzero F_π is also a necessary and sufficient condition of the chiral symmetry breaking. This energy scale also fits the most relevant NMS, the neutral-kaon system; the kaons having a mass of $\approx 0.5\,\mathrm{GeV}$.

The Lagrangian (6) is not expressed in terms of definite discrete symmetries. One that does can be constructed by taking the difference and sum of the $(a_L)^{\mu\alpha\beta}$ and $(a_R)^{\mu\alpha\beta}$ matrices. Keeping only the C-odd part, the quark-level Lagrangian for $d = 5$ becomes

$$
\delta\mathcal{L}^{a^{(5)}}_{\mathrm{qSME}} = \bar{\Psi}_A (a^{(5)})^{\mu\alpha\beta}_{AB} \gamma_\mu \partial_{(\alpha}\partial_{\beta)} \Psi_B, \tag{11}
$$

with a C-odd fermion term constructed from the left and right matrices in Eq. (6) added as

$$(a^{(5)})^{\mu\alpha\beta} = \frac{1}{2}\big((a_L)^{\mu\alpha\beta} + (a_R)^{\mu\alpha\beta}\big). \tag{12}$$

At the meson level, a three-index, suitable, free general Lagrangian using U given by Eq. (10) must be constructed such that it transforms the same under C, P, and T as the quark Lagrangian and has $\mathrm{SU(3)_L} \times \mathrm{SU(3)_R}$ symmetry. After examining the scalar effective Lagrangian found this way, once again restricting to C-odd free quadratic terms and isolating the term for neutral kaons the expression we are left with is

$$\mathcal{L}^{(5)}_{K_0} \supset \frac{1}{2} i\bar{K}_0 (k_a^{(5)})^{\mu\alpha\beta}_{K^0} \partial_\mu \partial_\alpha \partial_\beta K_0 + \text{h.c.} \tag{13}$$

For the connection between the scalar coefficient and the constituent strange- and down-quark coefficients the expression

$$(k_a^{(5)})^{\mu\alpha\beta}_{K^0} = r_8\big(a_s^{(5)\mu\alpha\beta} - a_d^{(5)\mu\alpha\beta}\big) \tag{14}$$

is found, where r_8 is some constant left undetermined by χPT.

5. Bounding $d = 5$ CPTV with Neutral-Kaon Experiments

The ξ experimental parameter of neutral-meson oscillations has been defined in Eq. (3) and is now considered as a source for bounding the above dimension-five nonminimal SME coefficients. As shown in Ref. 1, this ξ observable is generally boost- and orientation-dependent.

A significant part of finding estimates was done for the scalar coefficient components in Ref. 4. Since this work is within the restriction to light quarks, the most useful NMS is that of the neutral kaons with

$$\xi^{(5)}_{K^0} = \frac{m^2}{\Delta\lambda} \beta_\rho \beta_\sigma \beta_\tau k_a^{(5)\rho\sigma\tau} \tag{15}$$

being the meson-level expression of ξ arising from the scalar EFT coefficient $k_a^{(5)}$. Applying the connection (14), the quark-level expression for $\xi^{(5)}$ is therefore

$$\xi^{(5)}_{K^0} = \frac{m^2}{\Delta\lambda} \beta_\rho \beta_\sigma \beta_\tau r_8 \big(a_s^{(5)\rho\sigma\tau} - a_d^{(5)\rho\sigma\tau}\big). \tag{16}$$

One comment immediately obvious is that due to the two additional partial derivatives at $d = 5$, instead of the boost factor γ seen in the $d = 3$ minimal SME, an extra factor of $(\gamma m_K)^2$ arises in the expression for the observable $\xi^{(5)}$. For components with spatial indices, each spatial index also adds a factor of the magnitude of the velocity. This way of giving estimates from $d = 3$ measurements was already accomplished in Ref. 4 for the k_a coefficients using the factor $m^2\gamma^3$ for terms without spatial components.

The enhancement due to high boosts is widely recognized in NMS experiments and separates fixed target, high-boost experiments, such as E773 and KeTV,[1] from kaon-factory experiments where kaons are produced from the ϕ quarkonium nearly at rest.[11] The boost factor is an experiment-specific value. At KLOE, for instance, it is $\gamma = 1.03$. Knowing also the kaon mass $m_K \approx 0.5\,\mathrm{GeV}$, some coefficient estimates where geometric orientation factors are irrelevant or negligible can lead to $d = 5$ bounds from measurements targeting $d = 3$ minimal coefficients.[4]

Compared to a fixed spacetime-vector background, the sidereal rotation of the Earth allows tight bounds on the spatial components of the SME coefficients customarily marked with the X and Y Sun-Centered Celestial Equatorial Frame (SCCEF) indices. Again, moving from $d = 3$ to $d = 5$ the velocity factors also introduce higher-harmonic expressions in the sidereal rotation frequency. In particular, at $d = 5$ one must consider first-, second-, and third-harmonic variations. From earlier results at KLOE,[11] the first-harmonic analysis is readily available.

While the boost is low, one can take advantage of a full directional spread of the produced kaons and the large number of clean decays. This could then be adapted to $k_a^{(5)}$ of a single spatial index and carry it over to $a^{(5)}$ using the connection from χPT. Bounding higher harmonics asks for a suitable sidereal-time dependent analysis entailing a combination of binning and fitting sidereal variations over a sidereal day.

The CP-specific mode for decaying kaons is $\pi^+\pi^-$ while for the semileptonic decays, indicated above as SME phenomenology,[7] the flavor-specific modes $\pi^-e^+\nu$ and $\pi^+e^-\bar{\nu}$ are particularly beneficial.[12] The charge asymmetry of the lepton from a semileptonic decay of K_S or K_L can mark asymmetries between flavor-specific modes

$$A_{S,L} = \frac{\Gamma(K_{S,L} \to \pi^-e^+\nu) - \Gamma(K_{S,L} \to \pi^+e^-\bar{\nu})}{\Gamma(K_{S,L} \to \pi^-e^+\nu) + \Gamma(K_{S,L} \to \pi^+e^-\bar{\nu})}. \tag{17}$$

Assuming that $\Delta S = \Delta Q$ holds, the difference of A_S and A_L relates to ξ as $A_S - A_L = 2\Re(\xi)$.

The two-pion analysis entails recording decay intensities as a function of the decay time differences $\Delta\tau$ between the first and second decay of the two produced kaons. A distribution analysis of the intensities as a function of this time difference

$$A = \frac{I_{\pm}(\Delta\tau > 0) - I_{\pm}(\Delta\tau < 0)}{I_{\pm}(\Delta\tau > 0) + I_{\pm}(\Delta\tau < 0)} \tag{18}$$

gives a highly sensitive means to access CPTV, a feature of the destructive quantum interference at $\Delta\tau = 0$.[12]

At a kaon factory, the heart of a detailed analysis is proper binning of data and integrating over ranges of geometrical variables. The effect of these geometrical variables can be tracked if the kaon momentum and the background vector are transferred to the same SCCEF where their relative orientation can be described.[13]

In addition to the velocity magnitude, $\xi^{(5)}$ depends on the Earth's rotation, the detector coordinates, and the lab orientation. This introduces two constants, the colatitude and the angle of the colliding electron-positron beams in the East-South plane into the transformation equations between frames. For the KLOE experiment, one must also define two variable angles for the laboratory spherical coordinates ϕ and Θ. Since the kaons have back-to-back decays in any direction, this gives an advantage for orientation analyses as compared to mesons boosted into a narrower beam. It is especially useful in isolating components along the Earth's polar axis Z in the SCCEF frame, the one spatial component that is independent of the sidereal rotation. With enough Φ quarkonium momentum left during the production of the kaon pair, at KLOE the kaons of approximately equal magnitude but opposite direction can be designated as backward- and forward-scattered mesons from the sign of $\vec{p}_K \cdot \vec{\Phi}$.

The Z component can be isolated using the Θ lab coordinate in a measurement of forward and backward scattering differences for the long-lived kaons as $\Delta A_L \equiv A_L(\cos\Theta > 0) - A_L(\cos\Theta < 0)$ and averaging over the sidereal variation.[12] The dependence on Θ will be a new geometric function in the nonminimal case and further thought must be given to components with one, two, or three Z indices.

If, in turn, the time component is of interest, a symmetric integration over the polar angle Θ is in order averaging out the dependence on $\cos\Theta$. The integration is also necessary over sidereal time,[12] which from the still remaining periodicity will be straightforward for all harmonics.

An independent constraint can also be found for the purely time and Z components from the two-pion analysis.[12]

Mesons produced from quarkonium at threshold are quantum correlated with various consequences in NMS tests. Here, it is important to note that each correlated kaon must have its own experimental parameter designated, such as ξ_1 and ξ_2.

While in a correlated state, it is the sum and difference of these two ξ values that would represent a CPTV effect. This is the case between the time of production until the first meson decays. The probability amplitudes must reflect this from the time of production.[13] The difference in amplitudes due to momentum-dependent CPTV effects is specific to the SME analysis.

Asymmetries in the first and second decays in semileptonic and two-pion decays have been analyzed, as well as the time development of the surviving particle with its own phenomenological constant. These observable effects are providing the needed bounds. For $d = 5$ bounds, the component-by-component details as presented by the KLOE experiment for $\Delta a_\mu^{(3)}$ can be used with finding the right correspondences and by a modification based on the extra factors of momenta.

The analysis of unstable correlated mesons coupled to an LV background via a CPT-odd coupling merits further scrutiny. However, the Weisskopf–Wigner approximation is a suitable approach to properly include CPTV phenomena in the evaluation of correlated neutral-meson oscillations.

In summary, a more educated relationship between the SME coefficients of the pseudoscalar mesons and the coefficients of their constituent quarks can be found regarding the mesons as NG bosons in a spontaneous chiral symmetry breaking for the three light quarks. Neutral kaons, which have strange and down quark constituents, can then provide bounds for both sectors.

References

1. V.A. Kostelecký, *Phys. Rev. Lett.* **80**, 1818 (1998).
2. V.A. Kostelecký, *Phys. Rev. D* **61**, 016002 (2000).
3. V.A. Kostelecký and R. Potting, *Phys. Rev. D* **51**, 3923 (1995).
4. B.R. Edwards and V.A. Kostelecký, *Phys. Lett. B* **795**, 620 (2019).
5. S. Scherer and M.R. Schindler, *A Primer for Chiral Perturbation Theory, Lecture Notes in Physics* **830**, 76–82 (2012).
6. I.I. Bigi and A.I. Sanda, *CP Violation*, Cambridge University Press, Cambridge, 2009.
7. V.A. Kostelecký, *Phys. Rev. D* **64**, 076001 (2001).

8. O.W. Greenberg, *Phys. Rev. Lett.* **89**, 231602 (2002).
9. V.A. Kostelecký and Z. Li, *Phys. Rev. D* **99**, 056016 (2019).
10. R. Kamand, B. Altschul, and M.R. Schindler, *Phys. Rev. D* **95**, 056005 (2017); Phys. Rev. D **97**, 095027 (2018); B. Altschul and M.R. Schindler, *Phys. Rev. D* **100**, 075031 (2019).
11. KLOE-2 Collaboration, D. Babusci *et al.*, *Phys. Lett. B* **730**, 89 (2014).
12. A. Di Domenico, in V.A. Kostelecký, ed., *CPT and Lorentz Symmetry IV*, World Scientific, Singapore (2008).
13. A. Roberts, *Phys. Rev. D* **96**, 116015 (2017).

Chapter 8

Cosmic Backgrounds in the Gravitational Standard-Model Extension

C.M. Reyes,[*] C. Riquelme,[*,†] M. Schreck,[‡] and A. Soto,[§]

[*] *Centro de Ciencias Exactas, Departamento de Ciencias Básicas, Facultad de Ciencias, Universidad del Bío-Bío, Chillán, Casilla 447, Chile*

[†] *Departamento de Física, Universidad de Concepción, Casilla 160-C, Concepción, Chile*

[‡] *Departamento de Física, Universidade Federal do Maranhão, Campus Universitário do Bacanga, São Luís (MA), 65085-580, Brazil*

[§] *School of Mathematics, Statistics and Physics Newcastle University, Newcastle upon Tyne, NE1 7RU, UK*

We consider background fields within the gravitational sector of the Standard-Model Extension in a cosmological setting. Our analysis is divided into two parts. The first part addresses the consistency of nondynamical backgrounds in scenarios where diffeomorphism invariance is explicitly broken. Focusing on gravitational systems that admit Killing vector fields and possess *a priori* symmetries, we demonstrate that potential discrepancies between Riemannian geometry and dynamical equations can be avoided. The second part presents a direct application of various techniques developed by decomposing the modified Einstein equations along normal and tangential directions of the $(3+1)$ decomposition. We show that nondynamical backgrounds can lead to accelerated cosmological expansion without requiring a cosmological constant, thereby opening new avenues for interpreting dark energy.

1. Introduction

The cosmological Standard Model ΛCDM provides a remarkably precise description of the universe's evolution. It accurately accounts for the cosmic microwave background (CMB), primordial nucleosynthesis, and the formation of large-scale structures. Despite its success, the model faces several limitations, including the unknown nature of dark matter, the mechanism behind late-time accelerated expansion, and the lack of a fundamental description of physics in the pre-inflationary epoch.

Background fields in gravity have been proposed to unveil potential effects of diffeomorphism violation;[1] see also Ref. 2. Recently, both dynamical and nondynamical backgrounds have been considered as viable candidates for dark energy.[3–5] These developments are particularly compelling in light of recent DESI results,[6,7] which, when combined with CMB and supernovae data, suggest that accelerated expansion may be better explained by a field evolving in time rather than a cosmological constant.

2. Dynamical and Nondynamical Backgrounds

On the one hand, the gravitational sector of the minimal Standard-Model Extension (SME) with explicit diffeomorphism violation can be written in the form

$$S_{\mathrm{eSME}} = \int_{\mathcal{M}} \mathrm{d}^4x \, \frac{\sqrt{-g}}{2\kappa} \left(R + 2\Lambda + \mathcal{L}_{ust}\right) + S_{\partial\mathcal{M}} + S_{\mathrm{matter}}. \tag{1}$$

Here $\kappa = 8\pi G$, with G denoting Newton's gravitational constant, g is the determinant of the metric $g_{\mu\nu}$, and Λ the cosmological constant. The Lagrangian $\mathcal{L}_{ust}$ is usually decomposed into three parts,

$$\mathcal{L}_{ust} = -uR + s^{\mu\nu} R_{\mu\nu} + t^{\mu\nu\rho\sigma} R_{\mu\nu\rho\sigma}, \tag{2}$$

where u, $s^{\mu\nu}$, and $t^{\mu\nu\rho\sigma}$ are nondynamical background fields that couple to the Ricci scalar R, the Ricci tensor $R_{\mu\nu}$, and the Riemann tensor $R_{\mu\nu\rho\sigma}$, respectively. By construction, these backgrounds do not exhibit fluctuations, which is expressed mathematically by $\delta X = 0$ with $X \in \{u, s^{\mu\nu}, t^{\mu\nu\varrho\sigma}\}$. The piece S_{matter} accounts for a matter contribution, which will be specified later.

The action term $S_{\partial\mathcal{M}} = \sum_{i=u,s,t} S^{(i)}_{\partial\mathcal{M}}$ contains boundary terms to ensure a well-posed variational principle.[3,8] They read

$$S^{(u)}_{\partial\mathcal{M}} = -\oint_{\partial\mathcal{M}} \mathrm{d}^3y \, \varepsilon \frac{\sqrt{q}}{\kappa} uK, \tag{3a}$$

$$S^{(s)}_{\partial\mathcal{M}} = \oint_{\partial\mathcal{M}} \mathrm{d}^3y\, \varepsilon \frac{\sqrt{q}}{2\kappa} \left(s^{ab}K_{ab} - s^{\mathbf{nn}}K\right), \tag{3b}$$

$$S^{(t)}_{\partial\mathcal{M}} = \oint_{\partial\mathcal{M}} \mathrm{d}^3y\, \varepsilon \frac{\sqrt{q}}{\kappa}\, 2t^{\mathbf{n}a\mathbf{n}b}K_{ab}. \tag{3c}$$

Here, q_{ab} is the induced metric on the boundary hypersurface $\partial\mathcal{M}$, q its determinant, and $\varepsilon = n_\mu n^\mu$, where n_μ is the boundary normal evaluated at each point of $\partial\mathcal{M}$. We have introduced the extrinsic curvature K^{ab}, its trace K, and the notation $X^{\mathbf{n}} := n_\mu X^\mu$.

Varying the action of Eq. (1) leads to the modified Einstein field equations[9]

$$G_{\mu\nu} + \Lambda g_{\mu\nu} = (T^{Rstu})_{\mu\nu} + \kappa (T_m)_{\mu\nu}, \tag{4}$$

where $G_{\mu\nu} = R_{\mu\nu} - (R/2)g_{\mu\nu}$ is the Einstein tensor and $(T_m)_{\mu\nu}$ denotes the energy-momentum tensor of the matter content. The contribution from the background fields is represented by the second-rank tensor

$$\begin{aligned} 2(T^{Rstu})^{\mu\nu} &= -\nabla^\mu\nabla^\nu u - \nabla^\nu\nabla^\mu u + 2g^{\mu\nu}\nabla^2 u + 2uG^{\mu\nu} + s^{\alpha\beta}R_{\alpha\beta}g^{\mu\nu} \\ &\quad + \nabla_\alpha\nabla^\mu s^{\alpha\nu} + \nabla_\alpha\nabla^\nu s^{\alpha\mu} - \nabla^2 s^{\mu\nu} - g^{\mu\nu}\nabla_\alpha\nabla_\beta s^{\alpha\beta} \\ &\quad + t^{\alpha\beta\gamma\mu}R_{\alpha\beta\gamma}{}^{\nu} + t^{\alpha\beta\gamma\nu}R_{\alpha\beta\gamma}{}^{\mu} + t^{\alpha\beta\gamma\delta}R_{\alpha\beta\gamma\delta}g^{\mu\nu} \\ &\quad - 2\nabla_\alpha\nabla_\beta t^{\mu\alpha\nu\beta} - 2\nabla_\alpha\nabla_\beta t^{\nu\alpha\mu\beta}. \end{aligned} \tag{5}$$

On the other hand, spontaneous spacetime symmetry breaking is introduced through a dynamical field acquiring a spacetime-dependent vacuum expectation value.[3,10,11] A widely studied setting is the bumblebee model[10]:

$$S_{\text{bumb}} = S_B + S^{(\mathrm{B})}_{\partial\mathcal{M}} + S_{\text{matter}}, \tag{6}$$

where

$$S_B = \int_{\mathcal{M}} \mathrm{d}^4x\, \sqrt{-g}\left[\frac{1}{2\kappa}(R + \xi B^\mu B^\nu R_{\mu\nu}) - \frac{1}{4}B^{\mu\nu}B_{\mu\nu} - V(B_\mu B^\mu - b^2)\right], \tag{7a}$$

$$S^{(\mathrm{B})}_{\partial\mathcal{M}} = \oint_{\partial\mathcal{M}} \mathrm{d}^3y\, \varepsilon \frac{\sqrt{q}}{2\kappa}\left[2K + \xi\left(B^a B^b K_{ab} - \left(B^{\mathbf{n}}\right)^2 K\right)\right]. \tag{7b}$$

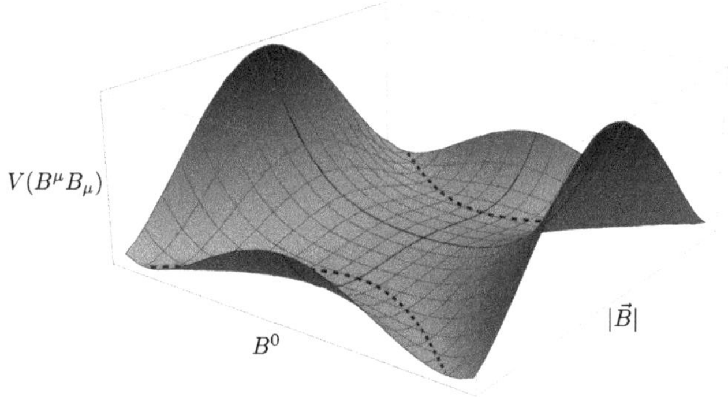

Fig. 1. Plot of the potential of the bumblebee field, which exhibits minima depicted by hyperbolas along the dashed lines. This structure arises from the spacetime metric signature and contrasts with the valley of minima in the Higgs potential, which is invariant under compact $SU(2)_L \otimes U(1)_Y$ transformations of the Higgs doublet.

We define the field strength tensor of the bumblebee field as $B_{\mu\nu} = \partial_\mu B_\nu - \partial_\nu B_\mu$ and assume that the bumblebee field acquires a vacuum expectation value, that is to say $\langle B_\mu \rangle = b_\mu$, in the potential

$$V(B_\mu B^\mu - b^2) = \frac{\lambda}{4}(B_\mu B^\mu - b^2)^2, \tag{8}$$

where λ is a coupling constant; see Fig. 1. The modified Einstein equations read

$$G^{\mu\nu} = (T_B)^{\mu\nu} + \kappa (T_m)^{\mu\nu}, \tag{9}$$

where

$$\begin{aligned}(T_B)^{\mu\nu} = \kappa\Big[&2V' B^\mu B^\nu + B^\mu{}_\kappa B^{\nu\kappa} - \Big(V + \frac{1}{4}B^{\lambda\kappa}B_{\lambda\kappa}\Big)g^{\mu\nu}\Big] \\ + \frac{\xi}{2}\Big[&B^\lambda B^\kappa R_{\lambda\kappa} g^{\mu\nu} - 2(g^{\mu\rho}B^\nu + g^{\nu\rho}B^\mu)B^\sigma R_{\rho\sigma} \\ &+ \nabla_\lambda \nabla^\mu (B^\lambda B^\nu) + \nabla_\lambda \nabla^\nu (B^\lambda B^\mu) \\ &- \nabla_\kappa \nabla_\lambda (B^\lambda B^\kappa) g^{\mu\nu} - \nabla_\lambda \nabla^\lambda (B^\mu B^\nu)\Big],\end{aligned} \tag{10}$$

and $(T_m)^{\mu\nu}$ is the energy-momentum tensor of matter. The equation of motion of the bumblebee field is

$$\nabla_\mu B^{\mu\nu} = 2V' B^\nu - \frac{\xi}{\kappa} B_\mu R^{\mu\nu}, \tag{11}$$

where the prime denotes differentiation with respect to the argument of V.

3. No-Go Constraint

Let us consider the following gravitational action with explicit breaking:

$$S = \int \mathrm{d}^4x \, \frac{\sqrt{-g}}{2\kappa} \left[R + \mathcal{L}'(g_{\mu\nu}, \bar{k}^{\alpha\beta\ldots\omega}) \right] + S_{\partial\mathcal{M}}, \tag{12}$$

where we have introduced the coefficients $\bar{k}^{\alpha\beta\ldots\omega}$ of a generic nondynamical background field. By construction, the action is defined to be invariant under observer diffeomorphisms, i.e., $\delta S_{\rm obs} = 0$. However, it breaks particle diffeomorphisms such that $\delta S_{\rm part} \neq 0$; see Ref. 12.

An observer transformation provides

$$\delta S_{\rm obs} = \int \mathrm{d}^4x \, \frac{\sqrt{-g}}{2\kappa} \left[\left(-G^{\mu\nu} + T'^{\mu\nu} \right) \delta g_{\mu\nu} + J_{\alpha\beta\ldots\omega} \delta\bar{k}^{\alpha\beta\ldots\omega} \right]. \tag{13}$$

Contrarily, a particle transformation leads to

$$\delta S_{\rm part} = \int \mathrm{d}^4x \, \frac{\sqrt{-g}}{2\kappa} \left[\left(-G^{\mu\nu} + T'^{\mu\nu} \right) \delta g_{\mu\nu} \right], \tag{14}$$

where we have defined

$$T'^{\mu\nu} =: \frac{1}{\sqrt{-g}} \frac{\delta(\sqrt{-g}\mathcal{L}')}{\delta g_{\mu\nu}}, \qquad J_{\alpha\beta\ldots\omega} =: \frac{\delta\mathcal{L}'}{\delta\bar{k}^{\alpha\beta\ldots\omega}}. \tag{15}$$

Performing several integrations by parts in Eq. (13) and employing the geometric identities $\nabla^\mu G_{\mu\nu} = 0$, we arrive at an identity for the background field coefficients:

$$\begin{aligned} 2\nabla_\mu T'^{\mu}{}_{\nu} &= J_{\alpha\beta\ldots\omega} \nabla_\nu \bar{k}^{\alpha\beta\ldots\omega} + \nabla_\lambda (J_{\nu\beta\ldots\omega} \bar{k}^{\lambda\beta\ldots\omega}) \\ &\quad + \nabla_\lambda (J_{\alpha\nu\ldots\omega} \bar{k}^{\alpha\lambda\ldots\omega}) + \cdots + \nabla_\lambda (J_{\alpha\beta\ldots\nu} \bar{k}^{\alpha\beta\ldots\lambda}). \end{aligned} \tag{16}$$

For the theory being dynamically consistent, a critical requirement is that the following differential equation be satisfied:[1,13]

$$\nabla_\mu T'^{\mu}{}_{\nu} = 0. \tag{17}$$

Equations (13) and (14) imply the intriguing integral equation

$$\delta S_{\rm part} + \int \mathrm{d}^4x \, \frac{\sqrt{-g}}{2\kappa} J_{\alpha\beta\ldots\omega} \delta\bar{k}^{\alpha\beta\ldots\omega} = \delta S_{\rm obs}. \tag{18}$$

In particular, an isometry generated by the Killing vector field χ satisfies $\delta\bar{k}^{\alpha\beta\ldots\omega} = \mathcal{L}_\chi \bar{k}^{\alpha\beta\ldots\omega} = 0$, where $\mathcal{L}_\chi$ is the Lie derivative along χ. Then, Eq. (18) implies $\delta S_{\rm obs} = 0 = \delta S_{\rm part}$, which restores[14] particle diffeomorphisms in the direction of χ.

In the following sections, we show that for a gravitational system that exhibits Killing vector fields, it is possible to fulfill Eq. (17). Alternatively, one may restrict the space of metric solutions of the modified Einstein equations by imposing extra conditions on spacetime geometry. For example, gravitational Chern–Simons theory[15] can be treated accordingly by requiring that ${}^*RR = 0$ with the dual Riemann tensor ${}^*R$.

4. Cosmology: Isotropy and Homogeneity

We start from a cosmological setting with the Friedmann metric

$$\mathrm{d}s^2 = -\mathrm{d}t^2 + a^2(t)\left[\frac{\mathrm{d}r^2}{1-kr^2} + r^2\left(\mathrm{d}\theta^2 + \sin^2\theta\mathrm{d}\phi^2\right)\right], \tag{19}$$

and consider the presence of a background which can be chosen to violate spacetime symmetries either explicitly or spontaneously. The modified Einstein equations are

$$G_{\mu\nu} + \Lambda g_{\mu\nu} = T_{\mu\nu} + \kappa(T_m)_{\mu\nu}, \tag{20}$$

where $T_{\mu\nu}$ depends on the background fields and $(T_m)_{\mu\nu}$ describes the energy-momentum of a perfect fluid.

Let us consider the transformations along the Killing directions χ characterizing isotropy and homogeneity and apply them to Eq. (20). Then,

$$\mathcal{L}_\chi T_{\mu\nu} = 0. \tag{21}$$

From the examples given in Ref. 3, we have found that in the specific sectors u, s, and t, the above condition leads to

$$\mathcal{L}_\chi u = \mathcal{L}_\chi s^{\mu\nu} = \mathcal{L}_\chi t^{\mu\nu\rho\sigma} = 0. \tag{22}$$

For spontaneous breaking in the bumblebee model, the Lie derivative of the bumblebee field is not restricted, but only its spatial norm. This procedure simplifies the differential equations of Eq. (17), which allows us to find nontrivial solutions of the backgrounds compatible with the dynamics.

5. Friedmann Equations and Cosmic Acceleration

This section is dedicated to applying the method of Killing vectors to both dynamical and nondynamical background fields. We focus on the t sector. Results on the u and s sectors including the Hubble tension can be consulted in Refs. 5 and 16.

5.1. *The t background*

The first step to analyze cosmology in the presence of backgrounds is to perform the ADM decomposition.[8,17,18] The decomposition of spacetime $\mathcal{M}(x^\mu)$ relies on the foliation in terms of a family of hypersurfaces $\Sigma_t(y^a)$ of constant time t. According to this method of decomposing spacetime, we define the projector $e^\mu_a := \frac{\partial x^\mu}{\partial y^a}$. The analysis simplifies when considering a purely tangential background t. Hence, we introduce the projections $t^{\mu\nu\rho\sigma} = e^\mu_a e^\nu_b e^\rho_c e^\sigma_d t^{abcd}$. After applying the method to decompose the modified Einstein equations, from Eq. (4) without cosmological constant, the first modified Friedmann equation can be written as

$$H^2 = \frac{1}{3\left(1 - \frac{1}{3} q_{ab} q_{cd} t^{acbd}\right)} \left(\kappa\rho - \frac{k}{a(t)^2} q_{ab} q_{cd} t^{acbd} - \frac{3k}{a(t)^2}\right), \tag{23}$$

and the second modified Friedmann equation reads

$$\begin{aligned} \dot{H} + H^2 = \frac{-1}{6\left(1 - \frac{1}{3} q_{ab} q_{cd} t^{acbd}\right)} \Bigg[& \kappa(\rho + 3P) + \frac{4k}{a(t)^2} q_{ab} q_{cd} t^{acbd} \\ & - 2q_{ac}\left(D_b D_d t^{abcd} + H q_{bd} \dot{t}^{abcd} + 2H^2 q_{bd} t^{abcd}\right)\Bigg], \end{aligned} \tag{24}$$

where $H(t) = \dot{a}(t)/a(t)$ is the Hubble parameter, D_a the induced covariant derivative on the hypersurface, and k the spatial curvature. The matter source is described by a perfect fluid of energy density ρ and pressure P with equation of state $P = w\rho$.

We follow the method previously explained to obtain a form-invariant background field; see Eq. (22). The result for spatially flat $k = 0$ is

$$t^{abcd} = a(t)^4 \eta \left(q^{ac} q^{bd} - q^{ad} q^{bc}\right), \tag{25}$$

with a constant η.

The deceleration parameter $q = -\ddot{a}a/\dot{a}^2$ is employed for the analysis. Configurations ought to be found that satisfy the condition $q < 0$. The latter is the case when $1 - 2\eta a^4 > 0$ and $(1+3w) - 2(5+3w)\eta a^4 < 0$, which can be summarized as

$$\left(\frac{1+3w}{5+3w}\right) \frac{1}{2\eta} < a^4 < \frac{1}{2\eta}. \tag{26}$$

We solve Eqs. (23) and (24) numerically by considering only matter ($w = 0$) to get an idea about the behavior of the scale factor. In this case, the relationship $\rho_0 = 3H_0\Omega_m/\kappa$ for the fluid density is valuable. It holds that

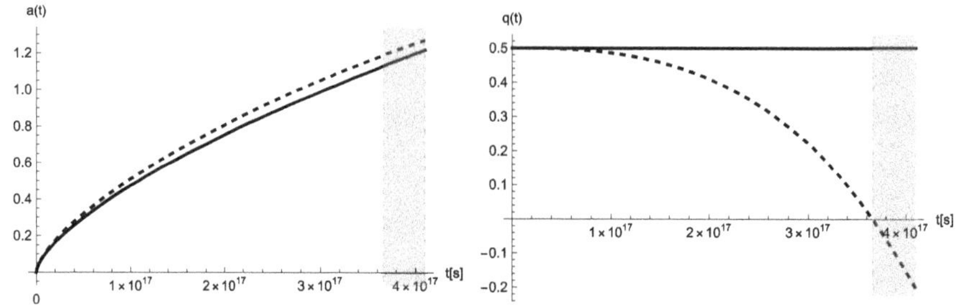

Fig. 2. Scale factor $a(t)$ and deceleration parameter $q(t)$ for the choice of $w = 0$. The solid lines show the standard behaviors and the dashed lines the solutions of the modified Friedmann equations (23) and (24). The gray areas illustrate the time intervals when accelerated expansion occurs.

$\Omega_m = 1$, since we analyze matter only. Moreover, the present value H_0 for the Hubble parameter determined by the Planck Collaboration[19] is chosen. As an example, $\eta = 5 \times 10^{-2}$ is taken, which leads to the plots for a and q displayed in the left- and right-hand-side panels, respectively, of Fig. 2. For the purpose of comparison, the standard behavior is presented, too.

5.2. *The bumblebee field*

For the bumblebee field with action (7a), we consider a purely tangential time-dependent bumblebee field

$$B^\alpha = e^\alpha_a B^a(t). \tag{27}$$

The modified Einstein equations (9) imply the first and second modified Friedmann equations. In particular,

$$H^2 = \frac{1}{3\left(1 - \frac{\xi}{3}B^c B_c\right)}\left[\kappa\rho + \kappa\left(V + \frac{1}{2}q^{cd}\dot{B}_c\dot{B}_d\right) - \frac{\xi}{2}\left(\frac{2k}{a(t)^2}B^c B_c + 2HB^c\dot{B}_c\right) - \frac{3k}{a(t)^2}\right], \tag{28a}$$

and

$$\dot{H} + H^2 = \frac{-1}{6\left(1 - \frac{\xi}{3}B^c B_c\right)}\left[\kappa(3P + \rho) + \kappa\left(2V' B^c B_c - 2V + q^{cd}\dot{B}_c\dot{B}_d\right) + \frac{\xi}{2}\left(-\frac{4k}{a(t)^2}B^c B_c + 4H^2 B^c B_c - 10Hq^{cd}\dot{B}_c B_d + 2q^{cd}\ddot{B}_c B_d + 2q^{cd}\dot{B}_c\dot{B}_d\right)\right]. \tag{28b}$$

The bumblebee field equations are

$$\ddot{B}_a + H\dot{B}_a + 2\left[V' - \frac{\xi}{2\kappa}\left(\frac{2k}{a(t)^2} + \dot{H} + 3H^2\right)\right]B_a = 0. \tag{29}$$

The central point of the above equations is that we have found possible configurations of the bumblebee field that respect the symmetries of isotropy and homogeneity by demanding $(T_B)^{\mu\nu}$ of the bumblebee field to fulfill Eq. (21). Notably, the arising configuration for the bumblebee field restricts the value of $B^a B_a$, but not the direction of B^a. Studying possible applications in this context is left for the future.

6. Conclusions

Here, we have compiled some of the critical findings of a series of recent works.[3,8,14] Scenarios of explicit diffeomorphism violation that avoid the no-go constraints have been presented. The modified Einstein equations were decomposed along the directions of the ADM decomposition to arrive at the modified Friedmann equations. Interestingly, regimes of accelerated cosmological expansion can arise without a cosmological constant. For the ADM decomposition of the bumblebee model and a more detailed discussion of cosmological phenomenology, the reader can consult Ref. 3.

Acknowledgments

CMR acknowledges support from Fondecyt 1241369. CR acknowledges support from the ANID fellowship No. 21211384 and Universidad de Concepción. MS is indebted to CNPq Produtividade 310076/2021-8 and CAPES/Finance Code 001.

References

1. V.A. Kostelecký, *Phys. Rev. D* **69**, 105009 (2004).
2. C. de Rham, G. Gabadadze, and A.J. Tolley, *Phys. Rev. Lett.* **106**, 231101 (2011).
3. C.M. Reyes, C. Riquelme, and A. Soto, *JCAP* **05**, 014 (2025).
4. N.A. Nilsson, *Phys. Dark Univ.* **45**, 101515 (2024).
5. C.M. Reyes, M. Schreck, and A. Soto, *Phys. Rev. D* **106**, 023524 (2022).
6. M. Abdul Karim *et al.* (DESI), *Phys. Rev. D* **112**, 083515 (2025).
7. K. Lodha *et al.* (DESI), *Phys. Rev. D* **112**, 083511 (2025).
8. C.M. Reyes and M. Schreck, *Phys. Rev. D* **104**, 124042 (2021).
9. Q.G. Bailey and V.A. Kostelecký, *Phys. Rev. D* **74**, 045001 (2006).
10. R. Bluhm and V.A. Kostelecký, *Phys. Rev. D* **71**, 065008 (2005).
11. J.V.V. Santos and M. Schreck, *Phys. Rev. D* **112**, 064026 (2025).

12. R. Bluhm, *Phys. Rev. D* **91**, 065034 (2015).
13. V.A. Kostelecký and Z. Li, *Phys. Rev. D* **103**, 024059 (2021).
14. C.M. Reyes, C. Riquelme, M. Schreck, and A. Soto, *Phys. Rev. D* **111**, 124011 (2025).
15. R. Jackiw and S.Y. Pi, *Phys. Rev. D* **68**, 104012 (2003).
16. M. Khodadi and M. Schreck, *Phys. Dark Univ.* **39**, 101170 (2023).
17. R.L. Arnowitt, S. Deser, and C.W. Misner, in *Gravitation: An Introduction to Current Research*, L. Witten (ed.) Wiley, New York, 1962.
18. K. O'Neal-Ault, Q.G. Bailey, and N.A. Nilsson, *Phys. Rev. D* **103**, 044010 (2021).
19. N. Aghanim *et al.* (Planck), *Astron. Astrophys.* **641**, A6 (2020) [Erratum: *Astron. Astrophys.* **652**, C4 (2021)].

Chapter 9

Full Single-Quantum Control of Particles in Penning Traps for Symmetry Tests at the Quantum Limit

J.M. Cornejo,[*,†] J.-A. Coenders,[*,†] A. Lissel,[†]
N. Poljakov,[*,†] M. Prasse,[*,†] Y. Priewich,[*,†] J. Schaper,[*,†]
M. Schubert,[‡] B. Hampel,[‡] M. Schilling,[‡]
S. Ulmer,[§,¶] and C. Ospelkaus[*,†,‖]

[*] *Institut für Quantenoptik, Leibniz Universität Hannover, Welfengarten 1, 30167 Hannover, Germany*
[†] *Laboratorium für Nano- und Quantenengineering, Leibniz Universität Hannover, Schneiderberg 39, 30167 Hannover, Germany*
[‡] *Institut für Elektrische Messtechnik und Grundlagen der Elektrotechnik, TU Braunschweig, Hans-Sommer-Straße 66, 38106 Braunschweig, Germany*
[§] *Heinrich-Heine-Universität Düsseldorf, Universitätsstraße 1, 40225 Düsseldorf, Germany*
[¶] *RIKEN, Ulmer Fundamental Symmetries Laboratory, 2-1 Hirosawa, Wako, Saitama, 351-0198, Japan*
[‖] *Physikalisch-Technische Bundesanstalt, Bundesallee 100, 38116 Braunschweig, Germany*

The BASE collaboration aims to measure antimatter systems with the highest precision in order to perform a rigorous test of CPT symmetry and search for physics beyond the Standard Model. As part of the BASE collaboration, we pursue the development of quantum logic inspired cooling and detection techniques for g-factor measurements of (anti-)protons. Implementing these methods requires full quantum-level control of individual antimatter particles confined in cryogenic Penning traps. By mapping the (anti-)proton's internal state onto a co-trapped $^9Be^+$ "logic" ion via free Coulomb coupling in a double-well potential, we can accelerate measurement cycles and push g-factor precision measurements on (anti-)protons toward the quantum limit. Here, we present an overview of the proposed method and the current status of the project, with special emphasis on the new cryogenic multi-Penning-trap stack and the proton detection system.

1. Introduction

The CPT theorem states that a relativistic Lorentz-symmetric quantum field theory must also be CPT invariant.[1] Although individual violations of C, P, or T symmetries are observed in certain weak interactions, the theorem guarantees that the combination of charge conjugation (C), parity inversion (P), and time reversal (T) remains an exact symmetry of all local, Lorentz-invariant quantum field theories.[2] This profound result not only underpins the consistency of particle–antiparticle equivalence but also provides critical constraints on possible extensions to the Standard Model.[3] Ultra-high precision comparisons of protons and antiprotons for the stringent test of CPT symmetry in the baryonic sector are pursued in the BASE (Baryon Antibaryon Symmetry Experiment) collaboration.[4] To achieve this, advanced cryogenic Penning-trap systems are used to measure the g-factor and the charge-to-mass ratio of protons and antiprotons to the highest precision. In this device, single particles are isolated in an ultra-high vacuum environment, where a strong magnetic field provides radial confinement, while static voltages applied to the trap electrodes ensure axial confinement. To date, the BASE collaboration has measured the g-factor with an unprecedented precision of 1.5 parts-per-billion (ppb) for the antiproton[5] and 0.3 ppb for the proton.[6] Additionally, it has carried out the most precise CPT test in the baryonic sector by comparing the proton and antiproton charge-to-mass ratios with a precision of 16 parts-per-trillion.[7] While the efficacy of these measurements is evident, current techniques continue to face challenges due to the finite temperature of the particles and the long preparation times that the measurement protocol entails. In this work, we report on the status of experimentally implementing quantum logic inspired cooling and detection techniques in order to boost current g-factor measurements of (anti-)protons within the BASE collaboration.

2. Quantum Logic Inspired Techniques for g-Factor Measurements with (Anti-)Protons

In recent decades, laser-based techniques have undergone substantial development for the manipulation of atomic ions.[8] The application of this concept to other quantum systems[9] has ultimately resulted in the experimental realization of quantum logic spectroscopy,[10] with application to molecular ions,[12] atomic clocks,[11] and highly charged ions.[13] Its use to measure the g-factor of (anti-)protons was introduced in Ref. 8 and it has been comprehensively discussed in Ref. 14. The main difference in its application to antimatter lies in the use of Penning traps instead of radio-frequency traps, which have been used for quantum logic spectroscopy so far.[10–13] Although recent years have seen growing interest in the application of laser-based techniques in Penning traps,[15–20] it has not been widely implemented in such devices due to the large Zeeman splittings of the electronic levels caused by the high magnetic fields,[21–23] as well as the complex cooling dynamics arising from the inherent instability of the magnetron motion.[24,25] This motion is one of the three characteristic motions of a particle in a Penning trap, along with the axial and reduced cyclotron motions. These three motions are related to the cyclotron motion through the invariance theorem.[26] The g-factor of an (anti-)proton in a Penning trap can be obtained by determining the cyclotron frequency ω_c and the spin-precession Larmor frequency ω_L according to

$$\frac{g}{2} = \frac{\omega_L}{\omega_c}. \tag{1}$$

The determination of the Larmor frequency of single (anti-)protons in a Penning trap using quantum logic spectroscopy involves a single $^9Be^+$ ion as a coolant and detection qubit. First, the $^9Be^+$ ion is cooled to its motional ground state via laser cooling and coupled to a single proton or antiproton through the free Coulomb interaction in a double-well potential.[8,9] When both ions oscillate at the same frequency, energy is exchanged between them.[27] Using this technique, a single proton or antiproton can be cooled to its motional ground state. Once both ions are initialized in their motional ground states, the spin state of the (anti-)proton is probed by applying different excitation frequencies around the expected Larmor frequency. For each interrogation, the spin state of the (anti-)proton is first transferred onto its motional mode via spin-motion coupling.[28] Depending on the spin orientation, a quantum of motion is added to the (anti-)proton, which is subsequently transferred to the $^9Be^+$ ion through the Coulomb interaction.

Finally, the motional state of the $^9Be^+$ ion is detected using laser sideband transitions, allowing the determination of the initial spin state of the (anti-)proton.

In order to implement this experimental approach, we have constructed and brought into operation a Penning trap system that includes a 5 T superconducting magnet and a cryogenic mechanical structure designed to provide laser access to the trap stack for manipulation of $^9Be^+$ ions.[29] Additionally, we have already demonstrated control of the cooling ion,[25,29] a novel laser system for Raman transitions,[22] sideband spectroscopy,[25] fast adiabatic transport,[30,31] and motional ground-state cooling of a single $^9Be^+$ ion.[32] The next step is to demonstrate energy exchange between pairs of $^9Be^+$ ions,[33] and subsequently with a $^9Be^+$ ion and a single proton.[14] For this, we have modified our cryogenic Penning trap system, as described in the following section.

3. Penning Trap Systems

Figure 1 presents a schematic of our Penning trap systems. This setup includes a beryllium trap for laser manipulation of single $^9Be^+$ ions, a coupling trap designed to study the coupling between two $^9Be^+$ ions, as a preparatory step for investigating the coupling between a $^9Be^+$ ion and a single proton, a precision trap for the manipulation of individual protons, a proton trap for the production of single protons, and a micro coupling trap specifically intended for coupling a $^9Be^+$ ion with a single proton. Several features have been improved compared to our previous setup (see Fig. 1), in which single-ion manipulation of $^9Be^+$ ions was already demonstrated. These improvements include modifications to the beryllium trap, the development of new coupling traps, including advancements in microfabrication techniques for the construction of the micro coupling trap, and upgrades to the precision and proton traps for proton production, initialization, and detection.

Ion-ion coupling via free Coulomb interaction is achieved in a specific region of our trap stack by generating a double-well potential. In this scenario, the energy exchange rate is given by

$$\Omega_{ex} = \frac{q_a q_b}{\pi \epsilon_0 s_0^3 \sqrt{m_a m_b}\sqrt{\omega_a \omega_b}}, \tag{2}$$

where s_0 is the distance between the ions, ϵ_0 is the vacuum permittivity, and m_i, q_i, and ω_i are the mass, charge, and oscillation frequency,

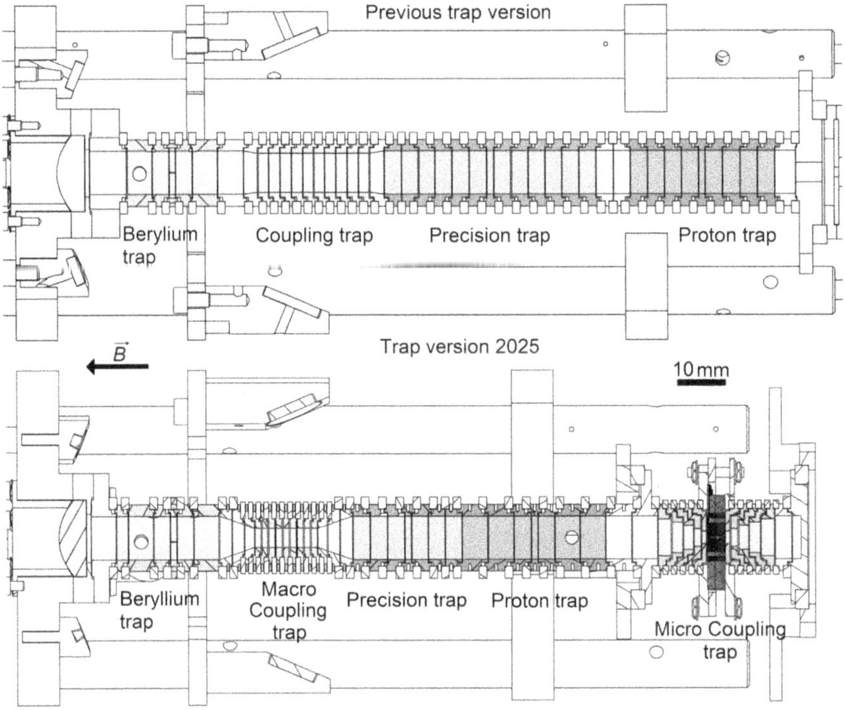

Fig. 1. Two-dimensional CAD drawing of the complete Penning trap stacks with specialized trap zones. The previous version of the trap is shown in the top panel. Several modifications have been implemented in the updated trap system to enable the manipulation of single protons. The electrodes are made of gold-plated oxygen-free high thermal conductivity (OFHC) copper and are electrically insulated by sapphire rings. The trap is located at the center of a 5 T superconducting magnet and cooled to a temperature of approximately 5 K. Further details are provided in the text.

respectively, of particle $i \in \{a, b\}$. In our previous Penning trap system,[25,31] several attempts were made to study energy exchange between pairs of $^9Be^+$ ions, albeit without success. We attribute this mainly to the lack of proper temperature initialization of the ions prior to each coupling attempt. This limitation arose from the inability to transport two different ions to the beryllium trap for laser manipulation because only the most recently produced ion could be transferred. In order to overcome this issue, an additional electrode was incorporated into the beryllium trap, enabling laser manipulation of two ions and reliable temperature initialization prior to each ion coupling attempt. Moreover, the four-segment ring electrode of the beryllium trap has been connected to independent DC voltage supplies in order to compensate for small deviations between the radial AC and DC potentials. With this improvement, we expect to Doppler-cool single $^{9}Be^{+}$ ions to the Doppler limit temperature of 500 μK, which is a factor of three lower than that achieved with our previous system.[25] This enhancement will facilitate ground-state cooling and reduce the overall cooling times.

In addition, we have introduced a smaller macro coupling trap. According to Eq. (2), a shorter distance between the ions, i.e., between the two potential minima generated in the double-well potential—leads to higher exchange rates. We have reduced the inner diameter of the trap by a factor of two, resulting in an eightfold increase in the coupling rate under the same conditions. This higher coupling rate facilitates energy exchange between two $^{9}Be^{+}$ ions and also enables the study of coupling between a single $^{9}Be^{+}$ ion and a proton. However, in the latter case, the diameter of the macro coupling trap is still too large to produce a double-well potential with a sufficiently small separation between the two potential minima, and thus a sufficiently large exchange rate. To overcome this limitation, a micro coupling trap has been developed, with an inner diameter reduced by a factor of five compared to the macro coupling trap. Under the same coupling conditions, this results in a 125-fold increase in the exchange rate. However, gold-plated copper electrodes cannot be used due to the small dimensions of this micro trap. Instead, it must be fabricated using microfabrication techniques, in a manner similar to that of planar surface radio-frequency ion traps[34] but adapted to a cylindrical geometry suitable for high-precision Penning trap experiments. This three-dimensional micro trap consists of 15 disc-shaped electrodes made of structured fused silica, partially gold-plated, with an outer diameter of 22 mm. It was designed as a self-aligning structure, where each electrode fits precisely into the next in only one specific orientation. The hollow cylindrical electrodes have an inner diameter of 800 μm and a thickness of 200 μm. Figure 2(a) shows a photo of a single electrode. Additionally, for proper ion transport through the micro trap, the magnetic field lines of the superconducting magnet must be aligned with the trap axis. To achieve this, we are currently developing an alignment system capable of slightly adjusting the position of the entire cryomechanical structure, allowing precise alignment of the trap axis with the magnetic field. Furthermore, several electrodes with progressively reduced inner diameters are used to guide the ions adiabatically into the coupling region, as shown in Fig. 1. Due to the novel and technically challenging task of transporting ions through the micro trap, in this first implementation, the micro trap has been positioned at one end of the trap stack. This configuration allows

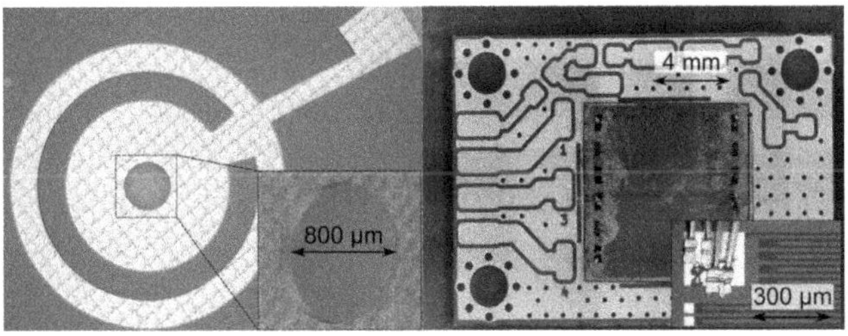

(a) (b)

Fig. 2. (a) Photograph of a single electrode from the micro coupling trap, showing the C-shaped structure designed for self-alignment. A detailed view of the inner hole used as a hollow cylindrical electrode, with an inner diameter of 800 µm, is shown in the bottom right corner. (b) Photograph of the printed circuit board (PCB) featuring four kinetic inductance resonators to be tested in our trap system. In the bottom right corner, a microscopic view of a meander structure with gold bonding is shown.

experiments with both protons and beryllium ions without obstructing the rest of the trap system. However, in future iterations, the micro trap will be placed between the beryllium trap and the precision trap to enable optimized coupling experiments.

For proton production, we plan to use an ablation laser and a tantalum target located at the proton trap, replacing the electron gun used in our previous setup. Tantalum is an effective adsorber of hydrogen, making it a suitable material for proton production via laser ablation.[35] Additionally, the precision trap has been shortened to accommodate the micro coupling trap and to avoid an excessive number of electrodes. For proton detection and thermalization, we will use axial and cyclotron resonators, as in our previous setup. These systems are based on resonant LC circuits, in which an inductor L is connected to a trap electrode. Due to the parasitic capacitance of the system, the circuit behaves as a resonant oscillator with a resonance frequency given by $\omega_{LC} = 1/\sqrt{LC}$. For the axial and cyclotron resonators, the resonance frequencies are matched to the axial and modified cyclotron oscillation frequencies of a single proton in the precision trap. This configuration allows both the detection of image currents induced in the trap electrodes by the proton and the thermalization of the proton to the temperature of the resonator. However, due to the significant difference in oscillation frequencies—approximately 600 kHz for the axial motion and 76 MHz for the reduced cyclotron motion—the two resonators require different designs. The axial resonator is made from superconducting NbTi cable, which provides a high quality factor Q. This enables the resonator to be placed at a considerable distance from the trap electrodes while maintaining sufficient sensitivity. In contrast, the cyclotron resonator is made from OFHC copper, which exhibits a lower Q value, requiring it to be located close to the trap electrodes for effective signal detection. In the updated trap design, this presents a challenge due to the physical size of the cyclotron resonator—several centimeters—as well as the need to accommodate mirrors and laser beam paths within the limited space. In order to overcome this limitation, we are also working on the implementation of a kinetic inductance-based resonator. Kinetic inductance arises from the inertial mass of mobile charge carriers in alternating electric fields, manifesting as an equivalent series inductance. This effect is particularly pronounced in conductors with high charge-carrier mobility, such as superconductors.[36] As inductance, we use thin films of high-temperature superconducting yttrium barium copper oxide ($YBa_2Cu_3O_{7-x}$, YBaCuO) over a $(LaAlO_3)_{0.3}(Sr_2TaAlO_6)_{0.7}$ (LSAT) substrate. Figure 2b shows the PCB with the substrate and the meander structure formed by the films. The YBaCuO thin films are deposited onto the LSAT substrate via pulsed laser deposition using polycrystalline targets.[38]

4. Summary and Outlook

In this work, we have discussed quantum logic inspired techniques for measuring the g-factor of (anti-)protons as a stringent test of CPT symmetry in the baryonic sector. Furthermore, we have presented an updated Penning trap system incorporating new methods for proton production and detection, as well as for implementing energy exchange between two $^9Be^+$ ions, and subsequently between a single $^9Be^+$ ion and a single proton. This is a key requirement for achieving quantum-level control of single protons. Alongside sideband spectroscopy of single protons, these represent the essential techniques to be developed. Additionally, due to advancements in the antiproton delivery system from CERN,[37] these methods could also be applied to antiprotons within the BASE collaboration.

Acknowledgments

This work was supported by PTB, LUH, and DFG through the clusters of excellence QUEST and QuantumFrontiers as well as through the Collaborative Research Center SFB1227 (DQ-mat Project-ID 274200144) and ERC StG "QLEDS." We also acknowledge financial support from the RIKEN

Pioneering Project Funding and the MPG/RIKEN/PTB Center for Time, Constants and Fundamental Symmetries.

References

1. W. Pauli, *Exclusion Principle, Lorentz Group and Reflexion of Space-Time and Charge,* p. 195, McGraw-Hill, New York, 1955.
2. R. Lehnert, *Symmetry* **8**, 114 (2016).
3. V.A. Kostelecký and N. Russell, 2025 edition, *Data Tables for Lorentz and CPT Violation,* arXiv:0801.0287v18.
4. C. Smorra *et al.*, *Eur. Phys. J. Spec. Top.* **224**, 3055 (2015).
5. C. Smorra *et al.*, *Nature* **550**, 371 (2017).
6. G. Schneider *et al.*, *Science* **358**, 1081 (2017).
7. M.J. Borchert *et al.*, *Nature* **601**, 53 (2022).
8. D.J. Wineland, C. Monroe, W.M. Itano, D. Leibfried, B.E. King, and D.M. Meekhof, *J. Res. Natl. Inst. Stand. Technol.* **103**, 259 (1998).
9. D.J. Heinzen and D.J. Wineland, *Phys. Rev. A* **42**, 2977 (1990).
10. P.O. Schmidt, T. Rosenband, C. Langer, W.M. Itano, J.C. Bergquist, and D.J. Wineland, *Science* **309**, 749 (2005).
11. T. Rosenband *et al.*, *Phys. Rev. Lett.* **104**, 070802 (2010).
12. F. Wolf, Y. Wan, J.C. Heip, F. Gebert, C. Shi, and P.O. Schmidt, *Nature* **530**, 457 (2016).
13. P. Micke *et al.*, *Nature* **578**, 60 (2020).
14. J.M. Cornejo *et al.*, *New J. Phys.* **23**, 073045 (2021).
15. J.F. Goodwin, G. Stutter, R.C. Thompson, and D.M. Segal, *Phys. Rev. Lett.* **122**, 143002 (2016).
16. S. Jain *et al.*, *Nature* **627**, 510 (2024).
17. E. Jordan *et al.*, *Phys. Rev. Lett.* **122**, 053603 (2019).
18. M. Bohman *et al.*, *Nature* **596**, 514 (2021).
19. F. Domínguez *et al.*, *J. Mod. Opt.* **65**, 613 (2018).
20. J.M. Cornejo *et al.*, *Int. J. Mass Spectrom.* **410**, 22 (2016).
21. M.J. Gutiérrez *et al.*, *New J. Phys.* **21**, 023023 (2019).
22. J. Mielke *et al.*, *J. Phys. B* **54**, 195402 (2021).
23. A.G. Paschke *et al.*, *Phys. Rev. Lett.* **122**, 123606 (2019).
24. R.C. Thompson and J. Papadimitriou, *J. Phys. B* **33**, 3393 (2000).
25. J.M. Cornejo *et al.*, *Phys. Rev. Res.* **5**, 033226 (2023).
26. L.S. Brown and G. Gabrielse, *Phys. Rev. A* **25**, 2423 (1982).
27. K.R. Brown, C. Ospelkaus, Y. Colombe, A.C. Wilson, D. Leibfried, and D.J. Wineland, *Nature* **471**, 196 (2011).
28. D. Nitzschke *et al.*, *Adv. Quantum Technol.* **3**, 1900133 (2020).
29. M. Niemann *et al.*, *Meas. Sci. Technol.* **31**, 035003 (2019).
30. T. Meiners *et al.*, *Eur. Phys. J. Plus.* **139**, 262 (2024).
31. M. von Boehn *et al.*, *Commun. Phys.* **8**, 107 (2025).
32. J.M. Cornejo *et al.*, *Phys. Rev. Res.* **6**, 033233 (2024).
33. T. Meiners *et al.*, *Hyperfine Interact.* **239**, 26 (2018).
34. S. Seidelin *et al.*, *Phys. Rev. Lett.* **96**, 253003 (2006).

35. L. Velardi, D. Delle Side, J. Krása, and V. Nassisi, *Nucl. Instr. Meth. B* **331**, 20 (2014).
36. R. Meservey and P.M. Tedrow, *J. Appl. Phys.* **40**, 2028 (1969).
37. M. Leonhardt *et al.*, *Nature* **641**, 871 (2025).
38. V.K. Malik *et al.*, *Phys. Rev. B* **85**, 054514 (2012).

Chapter 10

Supersymmetry with Lorentz Symmetry Violation

I.T. Drummond

Department of Applied Mathematics and Theoretical Physics,
Centre for Mathematical Sciences,
University of Cambridge, Wilberforce Road,
Cambridge, England

We investigate the effect that Lorentz symmetry violation can have on supersymmetry by considering two (massless free field) models, a photon/photino model and a Wess–Zumino model.

1. Introduction

We study two models that each exhibit Lorentz symmetry violation but retain the supersymmetry correspondence between the particles in each case.[1] The first model is a photon/photino model and the second is a Wess–Zumino model. We construct conserved supersymmetry currents and charges and investigate their algebraic properties and their effect on the quantum fields. The approach derives from developments in the Standard-Model Extension (SME).[2,3]

2. Models

The photon/photino model involves a vector field $A_\mu(x)$ and a Majorana spinor $\psi(x)$. The Lagrangian, $\mathcal{L}_V(x)$, for the photon is given by

$$\mathcal{L}_V(x) = -\frac{1}{4}F_{\mu\nu}F^{\mu\nu} + \frac{1}{8}C^{\mu\nu\sigma\tau}F_{\mu\nu}F_{\sigma\tau}, \tag{1}$$

where $F_{\mu\nu} = \partial_\mu A_\nu - \partial_\nu A_\mu$. The second term in Eq. (1) introduces Lorentz symmetry violation (LSV) by means of the Weyl tensor $C^{\mu\nu\sigma\tau}$ which is constructed from members of a null (Penrose) tetrad $\{l, n, m, \bar{m}\}$ satisfying $l^2 = n^2 = m^2 = \bar{m}^2 = 0$ and $l.n = -m \cdot \bar{m} = 1$. Remaining scalar products vanish. In our coordinate basis we have $l^\mu = (1,0,0,1)/\sqrt{2}$, $n^\mu = (1,0,0,-1)/\sqrt{2}$, $m^\mu = (0,1,i,0)/\sqrt{2}$ and $\bar{m}^\mu = (0,1,-i,0)/\sqrt{2}$. We have then

$$C^{\mu\nu\sigma\tau} = \kappa\left(A^{\mu\nu}A^{\sigma\tau} + \bar{A}^{\mu\nu}\bar{A}^{\sigma\tau}\right), \tag{2}$$

where $A^{\mu\nu} = l^\mu m^\nu - l^\nu m^\mu$ and $\bar{A}^{\mu\nu} = l^\mu \bar{m}^\nu - l^\nu \bar{m}^\mu$. Note that in Eq. (2) $C^{\mu\nu\sigma\tau}$ has the correct symmetry properties of a Weyl tensor. Note also

$$C^{\mu\nu\sigma\tau} l_\tau = 0. \tag{3}$$

Eq. (3) implies that l^μ is the unique principal null vector of the Weyl tensor which accordingly lies within Petrov class N. The parameter κ indicates the strength of the LSV.

2.1. *Photon dynamics*

From the Lagrangian $\mathcal{L}_V(x)$, we obtain the equation of motion

$$\partial_\mu(F^{\mu\nu} - C^{\mu\nu\sigma\tau}F_{\sigma\tau}) = 0. \tag{4}$$

On imposing the gauge condition $\partial.A = 0$, we obtain the wave equation

$$\partial^2 A^\nu - C^{\mu\nu\sigma\tau}\partial_\mu\partial_\sigma A_\tau = 0. \tag{5}$$

The plane wave solutions of Eq. (5) have the form $A_\nu = \varepsilon_\nu e^{-ip.x}$. In momentum space Eq. (5) takes the form

$$p^2\varepsilon^\nu - C^{\mu\nu\sigma\tau}p_\mu p_\sigma \varepsilon_\tau = 0. \tag{6}$$

There are two *physical* modes that provide wave functions for the propagating photons and two *unphysical* modes. These latter have the form $\varepsilon_\nu = p_\nu$ and $\varepsilon_\nu = l_\nu$ with a dispersion relation $p^2 = 0$. The unphysical modes travel on the *standard* lightcone. The dispersion relation for the physical modes has the form

$$(p^2)^2 - \kappa^2(l \cdot p)^4 = (p^2 + \kappa(l \cdot p)^2)(p^2 - \kappa(l \cdot p)^2) = 0. \tag{7}$$

Denoting the solutions as $p_\pm$, we have then

$$p_\pm^2 \pm \kappa(l \cdot p_\pm)^2 = 0. \tag{8}$$

The associated polarisation vectors are $\varepsilon_\pm^\mu$, where

$$\begin{aligned} \varepsilon_+^\mu &= l \cdot p_+ e_1^\mu - e_1 \cdot p_+ l^\mu, \\ \varepsilon_-^\mu &= l \cdot p_- e_2^\mu - e_2 \cdot p_- l^\mu. \end{aligned} \tag{9}$$

Here, e_1^μ and e_2^μ are coordinate 4-vectors along the 2- and 3-axes. For spatial momentum $\mathbf{p} = (p_1, p_2, p_3)$ the energies on the $\pm$ mass-shell cones are

$$E_\pm = \frac{1}{(1 \pm \kappa/2)}\left[\pm\frac{\kappa}{2}p_3 + \sqrt{\left(1 \pm \frac{\kappa}{2}\right)p_\perp^2 + p_3^2}\right], \tag{10}$$

where $p_\perp^2 = p_1^2 + p_2^2$. Under the interchange $\kappa \leftrightarrow -\kappa$ we have $E_+ \leftrightarrow E_-$, i.e., the mass-shell cones interchange. We assume $\kappa > 0$ and to avoid negative energies we will restrict $\kappa < 2$.

2.2. *Null-shear maps*

We introduce null-shear maps in momentum space. The maps are defined, for arbitrary 4-momentum p, by the formula(s)

$$p \to \hat{p}_\pm = p \pm \frac{\kappa}{2}(l.p)l. \tag{11}$$

The inverse map is

$$\hat{p}_\pm \to p = \hat{p}_\pm \mp \frac{\kappa}{2}(l \cdot \hat{p}_\pm)l. \tag{12}$$

It is obvious from its definition that $\hat{p}_\pm^2 = p^2 \pm \kappa(l \cdot p)^2$, with the implication that when p lies, say, on the $+$ mass-shell its image $\hat{p}_+$ lies on the standard shell and when p lies on the $-$ mass-shell $\hat{p}_-$ lies on the standard shell.

In configuration space, the appropriate definition of the null-shear is actually the inverse map. We have then

$$x \to \hat{x}_\pm = x \mp \frac{\kappa}{2}(l \cdot x)l. \tag{13}$$

If we set $\hat{\partial}_{\pm\mu} = \partial_\mu/\partial\hat{x}_{\pm\mu}$ then it follows immediately that

$$\hat{\partial}_{\pm\mu} = \partial_\mu \pm \frac{\kappa}{2}(l \cdot \partial)l_\mu. \tag{14}$$

Hence, ∂_μ and $\hat{\partial}_{\pm\mu}$ are related in the same way as p_μ and $\hat{p}_{\pm\mu}$. A further algebraic point is that $p \cdot x = \hat{p}_\pm \cdot \hat{x}_\pm$ with the result that

$$\hat{\partial}_\pm^2 e^{-ip\cdot x} = \hat{\partial}_\pm^2 e^{-i\hat{p}_\pm\cdot\hat{x}_\pm} = -\hat{p}_\pm^2 e^{-ip\cdot x} = 0, \tag{15}$$

when p lies on the appropriate mass-shell. This makes it clear that the lightcones in configuration space are indeed $x^2 \mp \kappa(l \cdot x)^2 = \hat{x}_{\pm}^2 = 0$. It is easy to verify that the three lightcones are nested, the + cone being the slow cone and the—cone being the fast supralightcone, the standard cone being sandwiched between those two and sharing with both, the vector l as a generator.

2.3. *Mode expansion for the vector field*

The quantum vector field can be expanded in terms of mode amplitudes $a_{\pm}(\mathbf{p})$ in the form

$$A_{\pm\mu}(x) = \int \frac{d^3\mathbf{p}}{(2\pi)^3} \frac{1}{\mathcal{N}_{\pm}(\mathbf{p})} \left[a_{\pm}(\mathbf{p}) e^{-ip_{\pm} \cdot x} + a_{\pm}^{\dagger}(\mathbf{p}) e^{ip_{\pm} \cdot x} \right] \varepsilon_{\pm\mu}, \tag{16}$$

where the mode operators satisfy the commutation relations

$$[a_{\pm}(\mathbf{p}), a_{\pm}^{\dagger}(\mathbf{p}')] = (2\pi)^3 \delta(\mathbf{p} - \mathbf{p}'). \tag{17}$$

Consistency with the canonical commutation relations requires

$$\mathcal{N}_{\pm}(\mathbf{p}) = \sqrt{(E_{\pm} - p_3)^2 (E_{\pm} \pm (\kappa/2)(E_{\pm} - p_3))}. \tag{18}$$

2.4. *Majorana spinor field*

In order to complete the photon/photino model, we require a Majorana spinor field $\psi(x)$ that satisfies a modified Dirac equation yielding a dispersion relation compatible with that for the vector field. The Lagrangian we choose has the form

$$\mathcal{L}_M(x) = \frac{i}{2} \bar{\psi}(x) \Gamma \cdot \partial \psi(x), \tag{19}$$

where

$$\Gamma^{\mu} = \gamma^{\mu} + \frac{1}{2} T^{\mu}_{\alpha\beta} \sigma^{\alpha\beta}, \tag{20}$$

and $\sigma^{\alpha\beta} = \frac{i}{2}[\gamma^{\alpha}, \gamma^{\beta}]$. We use the chiral representation for the γ-matrices. The Majorana equation is then

$$i\Gamma \cdot \partial \psi(x) = 0, \tag{21}$$

and we complete the model by setting $T^{\mu}_{\alpha\beta} = \xi l^{\mu} (A_{\alpha\beta} + \bar{A}_{\alpha\beta})$. The plane wave solutions have the form $\psi(x) = u e^{-ip \cdot x}$, with

$$(\Gamma \cdot p) u = 0. \tag{22}$$

For nontrivial solutions we require

$$\det(\Gamma \cdot p) = (p^2)^2 - 8\xi^2 (l \cdot p)^4 = 0, \tag{23}$$

This dispersion relation is compatible with that of the photon when $\xi = \pm\kappa/(2\sqrt{2})$. We will choose the $+$ sign from now on. The 4-spinor solutions of Eq. (22) are $u_\pm$ with entries $(-\bar{m} \cdot p_\pm, l \cdot p_\pm, \mp l \cdot p_\pm, \mp m \cdot p_\pm)$. It is useful to recognise that

$$u_\pm = \gamma \cdot \hat{p}_\pm u_\pm^{(0)}, \tag{24}$$

where $u^{(0)}$ is a 4-spinor with entries $(\mp i/\sqrt{2}, 0, 0, 1/\sqrt{2})$. It follows that

$$\gamma \cdot \hat{p}_\pm u_\pm = (\gamma \cdot \hat{p}_\pm)^2 u_\pm^{(0)} = (\hat{p}_\pm)^2 u_\pm^{(0)} = 0. \tag{25}$$

In configuration space the plane wave solutions of Eq. (21) are $\psi_\pm(x) = u_\pm e^{-ip_\pm \cdot x}$ and Eq. (25) implies that

$$i\gamma \cdot \hat{\partial}_\pm \psi_\pm(x) = 0, \tag{26}$$

and

$$\hat{\partial}_\pm^2 \psi_\pm(x) = 0, \tag{27}$$

thus confirming that the $\pm$ lightcones are indeed $\hat{x}_\pm^2 = 0$.

2.5. *Mode expansion for the Majorana field*

The quantum Majorana field can be expanded in terms of mode amplitudes $b_\pm(\mathbf{p})$ in the form

$$\psi_\pm(x) = \int \frac{d^3\mathbf{p}}{(2\pi)^3} \frac{1}{N_\pm(\mathbf{p})} \left[b_\pm(\mathbf{p}) e^{-ip_\pm \cdot x} \mp i b_\pm^\dagger(\mathbf{p}) e^{ip_\pm \cdot x} \right] u_\pm(p_\pm), \tag{28}$$

where

$$\{b_\pm(\mathbf{p}), b_\pm^\dagger(\mathbf{p}')\} = (2\pi)^3 \delta(\mathbf{p} - \mathbf{p}'). \tag{29}$$

The canonical anti-commutation relations are ensured by setting

$$N_\pm(\mathbf{p}) = \sqrt{2(E_\pm - p_3)(E_\pm \pm (\kappa/2)(E_\pm - p_3))}. \tag{30}$$

2.6. *Scalar fields*

In order to complete the Wess–Zumino model, we require scalar fields $\Phi_\pm(x)$ to partner the Majorana fields. The Lagrangian is

$$\mathcal{L}_S(x) = \frac{1}{2}(\hat{\partial}_+\Phi_+)^2 + \frac{1}{2}(\hat{\partial}_-\Phi_-)^2. \tag{31}$$

The equations of motion are

$$\hat{\partial}_\pm^2\Phi_\pm = 0. \tag{32}$$

The plane wave solutions are

$$\Phi_\pm(x) = e^{-ip_\pm\cdot x}, \tag{33}$$

with $\hat{p}_\pm^2 = 0$ yielding the same birefringence as the vector and Majorana fields. Supersymmetry relating the paired fields in the two models is clearly a possibility.

2.7. *Mode expansion for the scalar fields*

The scalar fields can be expanded in terms of mode amplitudes $c_\pm(\mathbf{p})$ in the form

$$\Phi_\pm(x) = \int \frac{d^3\mathbf{p}}{(2\pi)^3}\frac{1}{\mathcal{N}_{S\pm}(\mathbf{p})}\left[c_\pm(\mathbf{p})e^{-ip_\pm\cdot x} + c_\pm^\dagger(\mathbf{p})e^{ip_\pm\cdot x}\right], \tag{34}$$

where

$$[c_\pm(\mathbf{p}), c_\pm^\dagger(\mathbf{p}')] = (2\pi)^3\delta(\mathbf{p}-\mathbf{p}'), \tag{35}$$

and

$$\mathcal{N}_{S\pm}(\mathbf{p}) = \sqrt{2(E_\pm - p_3)(E_\pm \pm (\kappa/2)(E_\pm - p_3))}. \tag{36}$$

3. Supersymmetry charges

3.1. *Photon/photino model*

Using the null-shear maps described in Section (2.2), we set

$$\hat{A}_{\pm\mu} = A_{\pm\mu} \pm \frac{\kappa}{2}l_\mu l\cdot A_{\pm\mu} = A_{\pm\mu}. \tag{37}$$

The second equality arises because $l.A_\pm = 0$. We define a hatted version of the field strength

$$\hat{F}_{\pm\mu\nu} = \hat{\partial}_{\pm\mu}\hat{A}_{\pm\nu} - \hat{\partial}_{\pm\nu}\hat{A}_{\pm\mu}. \tag{38}$$

We can obtain hatted versions of the equation of motion

$$\hat{\partial}^{\mu}_{\pm}\hat{F}_{\pm\mu\nu} = 0, \tag{39}$$

the gauge condition

$$\hat{\partial}_{\pm} \cdot \hat{A}_{\pm} = 0, \tag{40}$$

and the Bianchi identity

$$\hat{\partial}_{\pm\lambda}\hat{F}_{\pm\mu\nu} + \hat{\partial}_{\pm\mu}\hat{F}_{\pm\nu\lambda} + \hat{\partial}_{\pm\nu}F_{\pm\lambda\mu} = 0. \tag{41}$$

We define hatted versions of the supercurrents

$$\hat{J}^{\mu}_{\pm\pm} = \hat{F}_{\pm\lambda\tau}\sigma^{\lambda\tau}\gamma^{\mu}\psi_{\pm}. \tag{42}$$

From Eq. (41) we can, by a standard argument, show that

$$\hat{\partial}_{\pm\mu}\hat{F}_{\pm\lambda\tau}\sigma^{\lambda\tau}\gamma^{\mu} = 0. \tag{43}$$

and from the equation of motion for $\psi_{\pm}$ the hatted conservation law

$$\hat{\partial}_{\pm\mu}\hat{J}^{\mu}_{\pm\pm} = 0. \tag{44}$$

Eq. (44) implies the unhatted version

$$\partial_{\mu}J^{\mu}_{\pm\pm} = 0, \tag{45}$$

where

$$J^{\mu}_{\pm\pm} = \hat{J}^{\mu}_{\pm\pm} \pm \frac{\kappa}{2}l^{\mu}l \cdot \hat{J}_{\pm\pm}. \tag{46}$$

We have then *conserved* supercharges

$$Q_{\pm\pm} = \int d^3\mathbf{x}J^0_{\pm\pm}(x). \tag{47}$$

Note that $Q_{\pm\mp}$ are *not* conserved (when $\xi = +\kappa/(2\sqrt{2})$). Using the mode expansions of the quantum fields, we obtain

$$Q_{++} = 2\int \frac{d^3\mathbf{p}}{(2\pi)^3}\frac{1}{\sqrt{E_+ - p_3}}\left[a_+(\mathbf{p})b^{\dagger}_+(\mathbf{p}) - ia^{\dagger}_+(\mathbf{p})b_+(\mathbf{p})\right]u_+(p_+), \tag{48}$$

and the algebraic result

$$\{Q_{++}, \bar{Q}_{++}\} = 4\int \frac{d^3\mathbf{p}}{(2\pi)^3}(a^{\dagger}_+(\mathbf{p})a_+(\mathbf{p}) + b^{\dagger}_+(\mathbf{p})b_+(\mathbf{p}))$$
$$\times \left[\hat{p}_+ \cdot \gamma + \frac{\varepsilon^{\lambda}_+(p_+)\hat{p}^{\tau}_+\sigma_{\lambda\tau}}{l.p_+}\right] \tag{49}$$

Note that the first term of Eq. (49) in square brackets is *similar* to the standard result and the second term is singular when $p \propto l$. There are similar results for Q_{--}.

3.2. *Wess–Zumino model*

In the Wess–Zumino model, we can construct a hatted version of the supercurrents in the form

$$\hat{J}^{\mu}_{\pm\pm}(x) = i\hat{\partial}_{\pm\mu}\Phi_{\pm}\gamma^{\lambda}\gamma^{\mu}\psi_{\pm}. \tag{50}$$

From the equations of motion for the fields, we have

$$\hat{\partial}_{\pm\mu}\hat{J}^{\mu}_{\pm\pm} = 0. \tag{51}$$

The unhatted current, constructed as before, is also conserved as are the charges $Q_{\pm\pm}$. In terms of the appropriate mode operators, we have

$$Q_{++} = \int \frac{d^3\mathbf{p}}{(2\pi)^3}\frac{1}{\sqrt{E_+ - p_3}}\left[-ic_+(\mathbf{p})b^{\dagger}_+(\mathbf{p}) - c^{\dagger}_+(\mathbf{p})b_+(\mathbf{p})\right]u_+(p_+). \tag{52}$$

The algebraic properties of $Q_{\pm\pm}$ are exactly, up to a normalisation, the same as in Eq. (49).

3.3. *Commutation relations of fields and supercharges*

It is simpler to exhibit the commutators of fields with supercharges for the Wess–Zumino model. Parallel results that are a little more complicated can be obtained in the photon/photino model. For the modes, we have

$$\begin{aligned} [Q_{++}, c_+(\mathbf{p})] &= \frac{1}{\sqrt{E_+ - p_3}} b_+(\mathbf{p})u_+(p_+) \\ \{b_+(\mathbf{p}), \bar{Q}_{++}\} &= -\frac{1}{\sqrt{E_+ - p_3}} c_+(\mathbf{p})\bar{u}_+(p_+). \end{aligned} \tag{53}$$

For the fields

$$\begin{aligned} [Q_{++}, \Phi_+(x)] &= \psi_+(x), \\ \{\psi_+(x), \bar{Q}_{++}\} &= \frac{-i}{2}\left[\gamma\cdot\hat{\partial}_+\Phi_+(x) - (e_{1\alpha}l_{\lambda} - e_{1,\lambda}l_{\alpha})\partial^{\alpha}\hat{\partial}_{+\tau}\sigma^{\lambda\tau}\mathcal{F}\right], \end{aligned} \tag{54}$$

where as can be checked directly from Eq. (31),

$$\mathcal{F} = -i\int \frac{d^3\mathbf{p}}{(2\pi)^3}\frac{1}{\mathcal{N}_{S+}(\mathbf{p})(l\cdot p)}(c_+(\mathbf{p})e^{-ip_+\cdot x} - c^{\dagger}_+(\mathbf{p})e^{ip_+\cdot x}). \tag{55}$$

This may be re-expressed in the form (the limit $\epsilon \to +0$ is implicit)

$$\mathcal{F} = \int_0^\infty ds e^{-\epsilon s}\Phi(x+sl). \tag{56}$$

The significant point here is that $\mathcal{F}$ in Eq. (55) is *nonlocal.* Thus, although the model has conserved supersymmetry operators it is not possible to deduce them by means of a standard Noether procedure. The same conclusion emerges from a detailed investigation of the photon/photino model. This remains true even when Lorentz invariance is restored. However, if we make the choice $\xi = -\kappa/(2\sqrt{2})$ then the matching of dispersion relations between the fields is switched round in such a way that we can obtain conserved charges $Q_{\pm\mp}$. Finally, setting $\kappa = 0$, we see that when Lorentz invariance is restored all four of the charges $Q_{\pm\pm}$ and $Q_{\pm\mp}$ are conserved. We can then construct an overall charge Q as

$$Q = Q_{++} + Q_{--} + Q_{+-} + Q_{-+}, \tag{57}$$

that has standard commutation relations and creates *local* variations in the fields in a conventional way. Non-local effects cancel.

4. Conclusions

Our conclusion is that, within the limitations of our models, supersymmetry of at least a partial kind can be maintained in the presence of Lorentz symmetry violation. Open questions remain, such as whether the constraints implied by this supersymmetry when masses and other interactions are included in the models and whether they are powerful enough to control the UV divergences that may then appear.

Acknowledgments

We acknowledge many helpful comments and suggestions by G.M. Shore and J.M. Drummond. This work has been partially supported by the United Kingdom Technologies Facilities Council Grant ST/X000664/1.

References

1. I.T. Drummond, *Phys. Rev. D* **108**, 045005 (2023).
2. M.S. Berger and V.A. Kostelecký, *Phys. Rev. D* **65**, 091701 (2002); S.G. Nibbelink and M. Pospelov, *Phys. Rev. Lett.* **94**, 081601 (2005); D. Colladay and P. McDonald, *Phys. Rev. D* **83**, 025021 (2011).

3. M. Schreck, *Phys. Rev. D* **89**, 085013 (2014); D. Colladay, P. McDonald, and R. Potting, *Phys. Rev. D* **89**, 085014 (2014); D. Colladay, P. McDonald, J. Noordmans, and R. Potting, *Phys. Rev. D* **95**, 025025 (2017); F.R. Klinkhamer and M. Schreck, *Phys. Rev. D* **78**, 085026 (2008); I.T. Drummond, *Phys. Rev. D* **95**, 025006 (2017).

Chapter 11

Hadronic Lorentz Violation in Chiral Perturbation Theory

M.R. Schindler

Department of Physics and Astronomy,
University of South Carolina,
Columbia, SC 29208, USA

Lorentz violation in hadronic systems is related to Lorentz-violating operators of quarks and gluons. Due to the nonperturbative nature of quantum chromodynamics (QCD) at low energies, establishing these relationships is complex. Chiral perturbation theory (ChPT) is an effective theory that provides one method of connecting quark- and gluon-level operators to those at the hadronic level, which can be used to calculate hadronic observables.

1. Introduction

Lorentz violation in the strongly-interacting sector of the Standard-Model Extension (SME)[1,2] is formulated in terms of quark and gluon degrees of freedom. On the other hand, experiments on strongly-interacting particles are performed on color-neutral hadrons. Because quantum chromodynamics (QCD) is nonperturbative at low energies, relating the properties and interactions of hadrons to those of the underlying quark and gluon degrees of freedom is challenging. This challenge already arises in the Lorentz-invariant conventional Standard Model context. Chiral perturbation theory (ChPT)[3–5] is an effective field theory of QCD, which is defined in terms of mesons and baryons. It is based on the (approximate) symmetries of QCD and is valid at energies well below 1 GeV. For a pedagogical introduction to ChPT, see, e.g., Ref. 6. ChPT has also been extended to incorporate the effects of Lorentz-violating (LV) quark and gluon operators in terms of hadronic operators. In the following I describe the basic principles behind this approach and give an example of how ChPT can help in setting new bounds on LV parameters.

2. Basics of Chiral Perturbation Theory

2.1. *Chiral symmetry*

Chiral perturbation theory is based on the symmetries of QCD. In addition to $\mathrm{SU}(3)_c$ color gauge symmetry and the discrete symmetries of parity (P), time reversal (T), and charge conjugation (C),[a] QCD exhibits an approximate chiral symmetry originating in the light quark masses. When considering processes at energies well below 1 GeV, the heavy quarks c, b, and t do not have to be included explicitly. The QCD Lagrangian for the light quark flavors is given by

$$\mathcal{L}_{QCD} = \sum_l \bar{q}_l (i\not{D} - m_l) q_l - \frac{1}{2}\mathrm{Tr}(\mathcal{G}^{\mu\nu}\mathcal{G}_{\mu\nu}), \tag{1}$$

where the index l denotes the light quark flavors, q_l are the light quark fields, and $\mathcal{G}_{\mu\nu}$ denotes the gluon field strength tensor. The following discussion is mainly restricted to up and down quarks ($l = u, d$), but the inclusion of the strange quark ($l = u, d, s$) is also possible. Because the light quark masses m_l are much smaller than typical hadronic scales, setting them to zero can serve as a useful theoretical starting point. This limit of $m_l \to 0$ is referred to as the chiral limit. It is convenient to collect the light quark flavors into a two-component multiplet $Q = (q_u, q_d)^T$ (or a three-component multiplet $(q_u, q_d, q_s)^T$ if the strange quark is included) and to introduce right- and left-handed quark fields,

$$Q_R = \frac{1}{2}(\mathbb{1} + \gamma_5)Q, \quad Q_L = \frac{1}{2}(\; - \gamma_5)Q, \quad \gamma_5 = i\gamma^0\gamma^1\gamma^2\gamma^3. \tag{2}$$

The QCD Lagrangian in the chiral limit then takes the form

$$\mathcal{L}^0_{QCD} = \bar{Q}_L i\not{D} Q_L + \bar{Q}_R i\not{D} Q_R - \frac{1}{2}\mathrm{Tr}(\mathcal{G}^{\mu\nu}\mathcal{G}_{\mu\nu}). \tag{3}$$

At the classical level, this Lagrangian is invariant under independent global U(2) transformations of the left- and right-handed fields,

$$Q_L \mapsto U_L Q_L, \quad Q_R \mapsto U_R Q_R, \tag{4}$$

[a]The QCD θ term is ignored in the following.

where $U_{L/R} \in \mathrm{U}(2)$, i.e., $\mathcal{L}^0_{QCD}$ possesses a classical $\mathrm{U}(2)_L \times \mathrm{U}(2)_R$ symmetry. Because of the axial anomaly, at the quantum level this symmetry reduces to an $\mathrm{SU}(2)_L \times \mathrm{SU}(2)_R \times \mathrm{U}(1)_V$ symmetry, where the $\mathrm{U}(1)_V$ symmetry is related to baryon number conservation and the invariance under $\mathrm{SU}(2)_L \times \mathrm{SU}(2)_R$ is referred to as chiral symmetry. One expected consequence of this chiral symmetry is the existence of degenerate hadron multiplets of opposite parity. The absence of this parity doubling combined with the observation that pions are much lighter than all other hadrons leads to the assumption that chiral symmetry is spontaneously broken to an $\mathrm{SU}(2)_V$ symmetry and that the pions are the Goldstone bosons of this spontaneous symmetry breaking.

In addition to the assumed spontaneous breaking of chiral symmetry, the finite light quark masses break chiral symmetry explicitly. The quark mass contribution to the Lagrangian of Eq. (1) takes the form

$$\mathcal{L}_M = -(\bar{Q}_R \mathcal{M} Q_L + \bar{Q}_L \mathcal{M}^\dagger Q_R), \tag{5}$$

where

$$\mathcal{M} = \begin{pmatrix} m_u & 0 \\ 0 & m_d \end{pmatrix} \tag{6}$$

is the mass matrix for the light quarks. Because $\mathcal{L}_M$ couples left- to right-handed fields, it is not invariant under chiral transformations. As a consequence of this explicit chiral symmetry breaking due to the quark masses, the pions as the Goldstone bosons of spontaneous chiral symmetry breaking are not massless.

2.2. *ChPT for mesons*

ChPT as the EFT of QCD at low energies incorporates the symmetries of QCD and their spontaneous and explicit breaking, but is formulated in terms of pions and nucleons instead of quarks and gluons. Chiral symmetry is realized nonlinearly on the pions,[3,7] which are collected in the SU(2) matrix

$$U(x) = \exp\left(i\frac{\phi(x)}{F}\right), \tag{7}$$

with

$$\phi(x) = \sum_{a=1}^{3} \phi_a \tau_a = \begin{pmatrix} \pi^0 & \sqrt{2}\pi^+ \\ \sqrt{2}\pi^- & \pi^0 \end{pmatrix}, \tag{8}$$

and the quantity F is identified with the pion decay constant in the chiral limit. Under chiral transformations $(L,R) \in \mathrm{SU}(2)_L \times \mathrm{SU}(2)_R$, the matrix of Eq. (7) transforms as

$$U(x) \mapsto R^\dagger U(x) L. \tag{9}$$

In the chiral limit, i.e., for zero quark masses, the ChPT Lagrangian is formulated in terms of the fields U and its derivatives. As long as one restricts the discussion to momenta much less than the scale $4\pi F \approx 1$ GeV, the effective Lagrangian can be expanded in the number of derivatives acting on U. The term with the lowest number of derivatives (aside from a constant) that is invariant under the assumed symmetries is given by

$$\frac{F^2}{4}\mathrm{Tr}\left[\partial_\mu U (\partial^\mu U)^\dagger\right]. \tag{10}$$

Expanding the matrix U in the pion fields leads not only to a canonical kinetic term for massless pions, but also to interaction terms with four and more pion fields. That all of these terms are related to a single low-energy coefficient (LEC) F is a consequence of chiral symmetry.

The inclusion of the quark mass term in the ChPT Lagrangian serves as the prototype for including LV effects. While the Lagrangian of Eq. (5) is not invariant under chiral transformations, it *would* be invariant *if*

$$\mathcal{M} \mapsto R\mathcal{M}L^\dagger. \tag{11}$$

The mass matrix $\mathcal{M}$, being a constant matrix, does not actually have this transformation behavior. However, by treating it as a spurion field with this behavior and constructing chirally invariant Lagrangians with it, the pattern of chiral symmetry breaking is mapped from the QCD term to the chiral Lagrangian. Imposing Lorentz invariance and invariance under discrete symmetries, the term with a single power of $\mathcal{M}$ is given by

$$\frac{F^2 B^2}{2}\mathrm{Tr}\left[\mathcal{M}U^\dagger + U\mathcal{M}^\dagger\right]. \tag{12}$$

The new LEC B is related to the scalar quark condensate and relates the quark masses to the lowest-order ChPT result for the pion masses, which assuming isospin symmetry $m_u = m_d = \hat{m}$ takes the form $M_{\pi,2}^2 = 2B\hat{m}$.

The complete leading-order (LO) ChPT Lagrangian for pions is

$$\mathcal{L}_2 = \frac{F^2}{4}\mathrm{Tr}\left[\partial_\mu U (\partial^\mu U)^\dagger\right] + \frac{F^2 B^2}{2}\mathrm{Tr}\left[\mathcal{M}U^\dagger + U\mathcal{M}^\dagger\right]. \tag{13}$$

ChPT is based on a simultaneous expansion of observables in derivatives and quark masses, divided by the breakdown scale Λ_b of the EFT. It is also possible to consider the coupling to external fields, such as electromagnetic or weak gauge fields. For this purpose it is convenient to promote the *global* chiral symmetry to a *local* symmetry.[4,5]

2.3. *ChPT for baryons*

Nucleons (and strange baryons) can also be treated in ChPT.[8] The proton (p) and neutron (n) are collected into a doublet $\psi = (p, n)^T$, which transforms under chiral transformations as

$$\psi \mapsto K(L, R, U)\psi, \tag{14}$$

where

$$K(L, R, U) = \sqrt{RUL^\dagger}^{-1} R\sqrt{U} \tag{15}$$

is an SU(2) matrix that not only depends on the left- and right-handed transformations L and R, but also on the pion fields through the matrix U. The lowest-order nucleonic Lagrangian is given by

$$\mathcal{L}_{\pi N}^{(1)} = \bar{\psi}\left(i\not{D} - m + \frac{g_A}{2}\gamma^\mu\gamma_5 u_\mu\right)\psi, \tag{16}$$

with the covariant derivative on the nucleon fields defined as

$$D_\mu\psi = (\partial_\mu - \Gamma_\mu)\,\psi, \quad \Gamma_\mu = \frac{1}{2}(u^\dagger\partial_\mu u + u\partial_\mu u^\dagger), \tag{17}$$

and the chiral vielbein

$$u_\mu = i\left(u^\dagger\partial_\mu u - u\partial_\mu u^\dagger\right). \tag{18}$$

The matrix u is the square root of the pion matrix U, i.e., $u^2 = U$. The Lagrangian of Eq. (16) contains two LECs: m, the nucleon mass in the chiral limit, and g_A, the nucleon axial-vector coupling constant, also taken in the chiral limit. Expanding u in terms of pion fields leads to a tower of pion-nucleon interaction terms. Unlike for pions, the nucleon mass in the chiral limit does *not* vanish in the chiral limit, which makes renormalization and power counting for nucleons more complex than in the mesonic sector.

3. Lorentz Violation in ChPT

The ChPT formalism has been applied to include the effects of LV operators from the minimal SME.[9–13] The following presents a representative example taken from Ref. 10 of how LV operators can be incorporated in ChPT; a more detailed discussion and applications to other operators, as well as the coupling to external fields, is found in the literature.

Consider the CPT-even, dimension-4 quark operators with LV parameters $c_{\mu\nu}$. After a naive restriction to two light flavors, the quark-level Lagrangian takes the form

$$\mathcal{L}_{\text{quark}}^{\text{LV}} = \frac{i}{2}\bar{Q}_L C_L^{\mu\nu}\gamma_\mu \overleftrightarrow{\mathcal{D}}_\nu Q_L + \frac{i}{2}\bar{Q}_R C_R^{\mu\nu}\gamma_\mu \overleftrightarrow{\mathcal{D}}_\nu Q_R, \tag{19}$$

where $\bar{Q}\overleftrightarrow{\mathcal{D}}_\nu Q$ is the covariant derivative acting on the quark fields and the LV parameters are collected in the matrices

$$C_{L/R}^{\mu\nu} = \begin{pmatrix} c_{u_{L/R}}^{\mu\nu} & 0 \\ 0 & c_{d_{L/R}}^{\mu\nu} \end{pmatrix}. \tag{20}$$

In addition to violating Lorentz invariance, $\mathcal{L}_{\text{quark}}^{\text{LV}}$ is *not* invariant under chiral transformations,

$$Q_L \mapsto LQ_L, \quad Q_R \mapsto RQ_R.$$

Similar to the case of a finite quark-mass matrix, the constant matrices $C_{L/R}^{\mu\nu}$ are promoted to spurion fields: the quark-level LV Lagrangian is invariant if the $C_{L/R}^{\mu\nu}$ are assumed to transform under chiral transformations as

$$C_L^{\mu\nu} \mapsto LC_L^{\mu\nu}L^\dagger, \quad C_R^{\mu\nu} \mapsto RC_R^{\mu\nu}R^\dagger. \tag{21}$$

It is also convenient to separate these matrices into their isoscalar and isovector components,

$$C_{L/R}^{\mu\nu} = {}^1C_{L/R}^{\mu\nu}\mathbb{1} + {}^3C_{L/R}^{\mu\nu}\tau_3.$$

With this assumed transformation behavior, we can now construct chirally invariant Lagrangians in terms of pion and nucleon fields that map Lorentz violation from the quark to the hadronic level. Assuming that the $C_{L/R}^{\mu\nu}$ are traceless and symmetric in the Lorentz indices, there is a single term at LO in the chiral power counting in the purely mesonic sector,

$$\mathcal{L}_\pi^{\text{LV,LO}} = \beta^{(1)}\frac{F^2}{4}\left({}^1C_{R\mu\nu} + {}^1C_{L\mu\nu}\right)\text{Tr}[(\partial^\mu U)^\dagger \partial^\nu U]. \tag{22}$$

The parameter $\beta^{(1)}$ is a new LEC that incorporates short-distance physics. It is dimensionless and, based on naive dimensional analysis, is expected to be $\mathcal{O}(1)$. A term containing ${}^3C_{L/R}^{\mu\nu}$ vanishes for symmetric $C_{L/R}^{\mu\nu}$.

The structure in the nucleonic sector is more complex. At LO, the Lagrangian is given by

$$\begin{aligned}\mathcal{L}_{\pi N}^{\text{LV,LO}} &= \alpha^{(1)}\bar{\Psi}\left[({}^3\widetilde{C}_R^{\mu\nu} + {}^3\widetilde{C}_L^{\mu\nu})(\gamma_\nu iD_\mu + \gamma_\mu iD_\nu)\right]\Psi \\ &+ \alpha^{(2)}\left({}^1C_R^{\mu\nu} + {}^1C_L^{\mu\nu}\right)\bar{\Psi}(\gamma_\nu iD_\mu + \gamma_\mu iD_\nu)\Psi \\ &+ \alpha^{(3)}\bar{\Psi}\left[({}^3\widetilde{C}_R^{\mu\nu} - {}^3\widetilde{C}_L^{\mu\nu})(\gamma_\nu\gamma^5 iD_\mu + \gamma_\mu\gamma^5 iD_\nu)\right]\Psi \\ &+ \alpha^{(4)}\left({}^1C_R^{\mu\nu} - {}^1C_L^{\mu\nu}\right)\bar{\Psi}(\gamma_\nu\gamma^5 iD_\mu + \gamma_\mu\gamma^5 iD_\nu)\Psi + \text{H.c.} \qquad (23)\end{aligned}$$

where

$${}^3\widetilde{C}_L^{\mu\nu} = u\,{}^3C_L^{\mu\nu}u^\dagger, \quad {}^3\widetilde{C}_R^{\mu\nu} = u^\dagger\,{}^3C_R^{\mu\nu}u.$$

The four independent LECs $\alpha^{(i)}$ are again expected to be $\mathcal{O}(1)$.

The LECs $\beta^{(1)}$ and $\alpha^{(i)}$ are all related to the underlying quark-level LV parameters $c_{u/d_{L/R}}^{\mu\nu}$. A determination of the effective hadronic LECs in terms of the quark-level parameters requires a nonperturbative QCD calculation, which in principle might be feasible using lattice QCD. But even in the absence of such a calculation the ChPT expressions are still useful. Expanding the matrices U and u in the number of pion fields gives the LV corrections to the pion and nucleon propagators as well as LV interaction vertices. For the pions, the propagator corrections take the form

$$\mathcal{L}_\pi^{\text{LV,LO},2\phi} = \frac{\beta^{(1)}}{4}(c_{u_L}^{\mu\nu} + c_{d_L}^{\mu\nu} + c_{u_R}^{\mu\nu} + c_{d_R}^{\mu\nu})\partial_\mu\phi_a\partial_\nu\phi_a. \qquad (24)$$

Comparison with the canonical form for a spin-0 particle,

$$\mathcal{L}_{\text{spin-0}}^{\text{LV}} = \frac{1}{2}k^{\mu\nu}\partial_\mu\phi_a\partial_\nu\phi_a, \qquad (25)$$

gives the ChPT result for the relationship between the meson- and quark-level parameters:

$$k_\pi^{\mu\nu} = \frac{\beta^{(1)}}{2}(c_{u_L}^{\mu\nu} + c_{u_R}^{\mu\nu} + c_{d_L}^{\mu\nu} + c_{d_R}^{\mu\nu}). \qquad (26)$$

Analogous steps lead to the ChPT expressions relating the proton and neutron $c_{p/n}^{\mu\nu}$ coefficients to the quark-level parameters; e.g., for the proton

$$c_p^{\mu\nu} = \frac{1}{2}\left[\alpha^{(1)} + \alpha^{(2)}\right](c_{u_L}^{\mu\nu} + c_{u_R}^{\mu\nu}) + \frac{1}{2}\left[-\alpha^{(1)} + \alpha^{(2)}\right](c_{d_L}^{\mu\nu} + c_{d_R}^{\mu\nu}). \qquad (27)$$

Summing the expressions for the proton and neutron yields a result that is proportional to the same linear combination of quark-level parameters as the pion coefficient in Eq. (26),

$$c_p^{\mu\nu} + c_n^{\mu\nu} \propto c_{u_L}^{\mu\nu} + c_{u_R}^{\mu\nu} + c_{d_L}^{\mu\nu} + c_{d_R}^{\mu\nu}. \tag{28}$$

Using the assumption that the ChPT LECs $\beta^{(1)}$ and $\alpha^{(i)}$ are all expected to be $\mathcal{O}(1)$, one can use existing bounds on the proton and neutron parameters to constrain the pion parameter $k_\pi^{\mu\nu}$. Even a conservative assumption about the relative sizes of $\beta^{(1)}$ and the $\alpha^{(i)}$ results in improvements on the pion parameters by several orders of magnitude.[10]

4. Conclusions

Chiral perturbation theory is the EFT of QCD for energies below 1 GeV, formulated in terms of color-neutral hadrons. It has been extended to include LV operators. ChPT establishes a connection between the LV operators of the SME, which are formulated in terms of quarks and gluons, and the hadronic-level operators that are used to analyze experiments. The ChPT approach is based on the approximate chiral symmetry of QCD and the pattern of chiral symmetry breaking that the SME operators exhibit. Chiral symmetry imposes relationships on operators with varying numbers of fields, e.g., relating corrections to the pion propagator to vertices with four and more pions. ChPT thus provides a very useful tool for understanding how quark-level LV operators manifest themselves at the hadronic level, and its use has the promise to establish many new bounds on Lorentz violation in strongly-interacting particles.

Acknowledgments

I thank my collaborators R. Kamand and B. Altschul. This material is based upon work supported by the U.S. Department of Energy, Office of Science, Office of Nuclear Physics, under Award Numbers DE-SC0010300 and DE-SC0019647.

References

1. D. Colladay and V.A. Kostelecký, *Phys. Rev. D* **55**, 6760 (1997).
2. D. Colladay and V.A. Kostelecký, *Phys. Rev. D* **58**, 116002 (1998).
3. S. Weinberg, *Physica A* **96**, no.1-2, 327 (1979).
4. J. Gasser and H. Leutwyler, *Ann. Phys.* **158**, 142 (1984).
5. J. Gasser and H. Leutwyler, *Nucl. Phys. B* **250**, 465 (1985).
6. S. Scherer and M.R. Schindler, *Lect. Notes Phys.* **830**, pp. 1–338 (2012).
7. S.R. Coleman, J. Wess, and B. Zumino, *Phys. Rev.* **177**, 2239 (1969).

8. J. Gasser, M.E. Sainio, and A. Švarc, *Nucl. Phys. B* **307**, 779 (1988).
9. J.P. Noordmans, J. de Vries, and R.G.E. Timmermans, *Phys. Rev. C* **94**, 025502 (2016).
10. R. Kamand, B. Altschul, and M.R. Schindler, *Phys. Rev. D* **95**, 056005 (2017).
11. J.P. Noordmans, *Phys. Rev. D* **95**, 075030 (2017).
12. R. Kamand, B. Altschul, and M.R. Schindler, *Phys. Rev. D* **97**, 095027 (2018).
13. B. Altschul and M.R. Schindler, *Phys. Rev. D* **100**, 075031 (2019)

Chapter 12

Lorentz and CPT Tests in Neutron and Storage-Ring EDM Experiments

Yunhua Ding

Department of Physics and Astronomy, Ohio Wesleyan University, Delaware, OH 43015, USA

We investigate Lorentz- and CPT-violating effects in neutron and storage-ring electric dipole moment (EDM) experiments within the framework of the Standard-Model Extension (SME). For neutron EDM experiments, perturbation theory is applied to derive leading-order contributions to the spin-precession frequency arising from Lorentz and CPT violation. For storage-ring experiments, a generalized Bargmann–Michel–Telegdi equation is used to determine the corresponding spin-precession modifications. The analysis establishes explicit correspondences between measured EDMs and specific SME coefficients, providing a basis for setting the first limits on several previously unconstrained coefficients for Lorentz violation in future studies.

1. Introduction

Electric dipole moment experiments provide one of the most sensitive probes of possible new sources of P and CP violation in nature. For example, the neutron EDM (nEDM) has been measured with a precision of $|d_n| < 1.8 \times 10^{-26}\, e\,\mathrm{cm}$ (90% C.L.).[1] Measurements of the muon EDM (μEDM) yield $|d_{\mu^+}| < 2.1 \times 10^{-19}\, e\,\mathrm{cm}$ for antimuons and $|d_{\mu^-}| < 1.5 \times 10^{-19}\, e\,\mathrm{cm}$ for muons (95% C.L.).[2] The impressive sensitivities achieved in these measurements also make EDM experiments powerful tests of other fundamental symmetries, including Lorentz and CPT symmetry.

It has been shown that tiny violations of Lorentz and CPT symmetry can arise in certain candidate theories unifying gravity with quantum physics, such as string theory.[3] A comprehensive framework to describe all possible Lorentz-violating effects within effective field theory is provided by the Standard-Model Extension,[4] constructed by adding all possible Lorentz-violating terms into the action of General Relativity and the Standard Model. Each Lorentz-violating term involves a coordinate-independent contraction of a Lorentz-violating operator with a corresponding controlling coefficient. In effective field theory, any CPT violation is necessarily accompanied by Lorentz violation, so the SME also encompasses all possible CPT-violating effects.[5] Extensive high-precision tests across a wide range of subfields have placed stringent limits on many SME coefficients.[6]

In this work, we analyze Lorentz- and CPT-violating contributions to spin precession in both neutron and storage-ring EDM experiments within the SME framework. For confined neutrons, perturbation theory is used to determine the leading-order effects. For charged particles in storage rings, we apply the generalized Bargmann–Michel–Telegdi (BMT) equation to extract the corresponding spin-precession modifications.[7] We establish explicit relations between measured EDMs and specific SME coefficients. This enables future studies to place first constraints on various SME coefficients using current EDM limits reported by neutron and storage-ring experiments.[1,2]

2. Neutron EDM

In conventional Lorentz-invariant analyses of EDM effects involving confined neutrons, the primary observable is the shift in the neutron spin-precession frequency, corresponding to the energy difference between spin-up and spin-down states. To isolate the EDM contribution, experiments measure the spin-precession frequency with parallel and antiparallel electric and magnetic fields. The resulting difference is proportional to the nEDM according to

$$\Delta\omega_{n,\mathrm{EDM}} = 4d_n E, \tag{1}$$

where d_n is the nEDM and E is the applied electric-field strength.

Lorentz violation can induce frequency shifts even when the nEDM vanishes. The dominant energy shifts for neutrons in this case can be obtained via perturbation theory as $\delta E_s = \langle \chi_s | \delta \mathcal{H} | \chi_s \rangle$, where χ_s are unperturbed stationary neutron eigenstates with $s = \pm 1$ denoting spin orientation relative to the magnetic field, and $\delta\mathcal{H}$ is the perturbative Hamiltonian for Lorentz violation.[8] For convenience, we adopt an apparatus frame with magnetic field $\boldsymbol{B} = B\hat{x}_3$ and electric field $\boldsymbol{E} = E\hat{x}_3$. In this frame, reversing the electric-field direction while holding the magnetic field fixed yields a Lorentz-violating frequency difference

$$\Delta\omega_{n,\mathrm{LV}} = 4\widetilde{b}^{303}_{F,n} E, \tag{2}$$

where the tilde coefficient $\widetilde{b}^{303}_{F,n}$ is

$$\widetilde{b}^{303}_{F,n} = b^{(5)303}_{F,n} + H^{(5)1203}_{F,n} - m_n d^{(6)3003}_{F,n} - m_n g^{(6)12003}_{F,n}. \tag{3}$$

Comparison of Eqs. (1) and (2) reveals the correspondence

$$|d_n| \longleftrightarrow |\widetilde{b}^{303}_{F,n}|, \tag{4}$$

relating the measured nEDM to the relevant coefficients for Lorentz violation.

3. Storage-Ring EDM

Precision EDM measurements of charged particles, such as protons, muons, and their antiparticles, are performed in storage-ring experiments, where particles are confined by electric and/or magnetic fields, and the EDM-induced spin precession is measured via polarimetry from spin-dependent elastic scattering along the ring. The spin-precession frequency in the presence of an EDM is described by the BMT equation.[9] For a radially outward electric field $\vec{E}$ and a uniform upward magnetic field $\vec{B}$, the lab-frame EDM contribution to the spin-precession frequency for a particle of species w, EDM d_w, and velocity $\vec{\beta}$ is

$$\delta\vec{\omega}_{s,\mathrm{EDM}} = -2d_w\big(\vec{E} + \vec{\beta} \times \vec{B}\big). \tag{5}$$

Recently, the BMT equation was generalized to incorporate Lorentz and CPT violation.[7] From the generalized BMT equation, the modification to the radial component of the spin-precession frequency is

$$\delta\omega^{\mathrm{rad}}_{s,\mathrm{LV}} = \check{H}^{03}_{w} + E\big(\check{b}^{11}_{F,w} + \check{b}^{22}_{F,w} + \check{g}^{33}_{F,w}\big) + B\check{H}^{33}_{F,w}, \tag{6}$$

where the hachek coefficients are defined as

$$\begin{aligned}
\check{H}^{03}_{w} &= -\beta\Big(2H^{03} - m_w(d^{12} - d^{21}) - \gamma m_w(2g^{030} + g^{131} + g^{232})\Big),\\
\check{b}^{11}_{F,w} &= \tfrac{1}{\gamma}b^{101}_F + H^{2301}_F - m_w d^{1001}_F - \gamma m_w g^{23001}_F,\\
\check{b}^{22}_{F,w} &= \tfrac{1}{\gamma}b^{202}_F - H^{1302}_F - m_w d^{2002}_F + \gamma m_w g^{13002}_F,\\
\check{g}^{33}_{F,w} &= \gamma\beta^2 m_w\big(g^{03102}_F - g^{03201}_F\big),\\
\check{H}^{33}_{F,w} &= 2\beta H^{0312}_F - \beta m_w(d^{1212}_F - d^{2112}_F)\\
&\quad - \gamma\beta m_w\big(2g^{03012}_F + g^{13112}_F + g^{23212}_F\big),
\end{aligned} \tag{7}$$

with m_w and $\gamma = 1/\sqrt{1-\beta^2}$ indicating the particle's mass and Lorentz factor, respectively. Comparison of Eqs. (5) and (6) yields the correspondence between a measured EDM and the relevant SME coefficients,

$$|d_w| \longleftrightarrow \frac{|\delta\omega^{\rm rad}_{s,{\rm LV}}|}{2\,|\vec{E} + \vec{\beta}\times\vec{B}|}. \tag{8}$$

4. Summary

The relations in Eqs. (4) and (8) establish direct correspondences between EDM observables and the relevant SME coefficients for Lorentz violation in both neutron and storage-ring experiments. Placing bounds on the relevant SME coefficients in the Sun-centered frame requires transformations from the apparatus frame to the Sun-centered frame, as discussed in detail elsewhere.[8] Existing experimental limits on the neutron EDM and on the EDMs of muons and antimuons[1,2] can therefore be translated into new constraints on a variety of previously unbounded SME coefficients, an analysis that can be performed in future work.

Acknowledgments

This work is supported in part by the Teaching, Learning, and Innovation grant from Ohio Wesleyan University.

References

1. C. Abel *et al.*, *Phys. Rev. Lett.* **124**, 081803 (2020).
2. G.W. Bennett *et al.*, *Phys. Rev. D* **80**, 052008 (2009).
3. V.A. Kostelecký and S. Samuel, *Phys. Rev. D* **39**, 683 (1989); V.A. Kostelecký and R. Potting, *Nucl. Phys. B* **359**, 545 (1991); *Phys. Rev. D* **51**, 3923 (1995).

4. D. Colladay and V.A. Kostelecký, *Phys. Rev. D* **55**, 6760 (1997); *Phys. Rev. D* **58**, 116002 (1998); V.A. Kostelecký, *Phys. Rev. D* **69**, 105009 (2004).
5. O.W. Greenberg, *Phys. Rev. Lett.* **89**, 231602 (2002).
6. V.A. Kostelecký and N. Russell, *Data Tables for Lorentz and CPT Violation*, 2025 edition, arXiv:0801.0287v18.
7. Y. Ding, V.A. Kostelecký, and A.J. Vargas, *Phys. Lett. B* **867**, 139604 (2025).
8. Y. Ding and V.A. Kostelecký, *Phys. Rev. D* **94**, 056008 (2016).
9. V. Bargmann, L. Michel, and V.L. Telegdi, *Phys. Rev. Lett.* **2**, 435 (1959).

Chapter 13

Expanding the Reach of Laboratory SME Searches Using Higher-Precision Boost Transformations

Jay D. Tasson

Physics and Astronomy Department, Carleton College, Northfield, MN 55057, USA

Additional sensitivities to Lorentz violation can be obtained from existing experiments by considering additional boost-suppressed effects. The additional Lorentz-violating signals arise as variations in experimental observables at the commonly used sidereal frequency as well as more novel frequencies. In this work we provide some examples that serve to illustrate how interesting signals arise from the structure of the relevant boost transformations.

1. Introduction

The Standard-Model Extension (SME) is a test framework for organizing a systematic search for Lorentz and CPT violation, perhaps arising from Planck-scale physics.[1,2] Searches in the context of the SME have now been performed across nearly all areas of physics[3] ranging from atomic physics,[4] to gravitational waves,[5] to collider physics.[6] Due to the ubiquity of ordinary matter, searches with fermions such as protons, neutrons, and electrons have been a particularly active area. The complete minimal fermion sector of the SME (operators of mass dimension $d = 3$ and $d = 4$) was written down in 1998 by Colladay and Kostelecký.[1] Since then, dozens of experiments have sought Lorentz- and CPT-violating effects using fermions.[3] Despite more than 25 years of intense activity, even the coefficients associated with protons, neutrons, and electrons had not been completely explored as of the publishing of the recent edition of the data tables.[3] Our recent work illustrates that these unconstrained coefficients can be explored using higher-precision frame transformations in ways that both generate new constraints from existing published results and lead to new discovery potential for future analyses.[7] While assistance from computer algebra is needed to address the full space of 44 degrees of freedom per fermion that exists in the applicable limit of the minimal SME, it is instructive to examine simple examples of the general procedure that can be presented with a few lines of algebra. We present such examples in the remainder of this work that illustrate key features of our general results.

2. The Fermion Sector

We begin by presenting the parts of the Lagrange density for the minimal fermion sector that are relevant for our study as a means of defining the key variables for the analysis to follow.[1]

$$\mathcal{L} = \tfrac{1}{2} i\overline{\psi}(\gamma_\nu + d_{\mu\nu}\gamma_5\gamma^\mu + \tfrac{1}{2} g_{\lambda\mu\nu}\sigma^{\lambda\mu}) \overleftrightarrow{\partial^\nu} \psi - \overline{\psi}(m + b_\mu\gamma_5\gamma^\mu + \tfrac{1}{2} H_{\mu\nu}\sigma^{\mu\nu})\psi. \quad (1)$$

The objects b_μ, $d_{\mu\nu}$, $g_{\lambda\mu\nu}$, and $H_{\mu\nu}$ are the relevant coefficients for Lorentz violation. The coefficients a_μ, e_μ, and f_μ are not explicitly included in Eq. (1) as they can be removed from the theory in the limit that we consider in this work.[1,8] We avoid explicitly introducing the coefficient $c_{\mu\nu}$ because all of its components have been probed already by other work.[3] The trace of $d_{\mu\nu}$ is not Lorentz violating and is also not considered here. The fermion field is denoted ψ, γ^μ are the Dirac matrices, and m is the fermion mass. In the single-fermion limit here in flat spacetime, there are 44 observable degrees of freedom that are characterized by combinations of the coefficients appearing in the Lagrange density. Those 44 combinations are conveniently parameterized by the so-called tilde coefficients (see Table P56 of Ref. 3).

Some of the most sensitive experiments in the fermion sector[4,9,10] are those that look for spin-precession effects stemming from an interaction described by the Hamiltonian[11]

$$H = \tilde{b}_j \sigma_j, \quad (2)$$

where the $\tilde{b}_j$, three of the 44 tilde coefficients, are the combination of SME coefficients

$$\tilde{b}_j = b_j - \tfrac{1}{2}\varepsilon_{jkl} H_{kl} - m(d_{jt} - \tfrac{1}{2}\varepsilon_{jkl} g_{klt}), \quad (3)$$

and σ_j are the Pauli spin matrices. The SME coefficients are typically taken as spacetime constants in a Sun-centered coordinate system,[3] making them time-dependent in a laboratory frame on Earth due to the rotation of the Earth at the sidereal frequency ω, the boost of the Earth as it orbits the Sun ($\beta_\oplus \sim 10^{-4}$) at the annual frequency Ω, and the boost of the laboratory as it revolves around the Earth ($\beta_L \lesssim 10^{-6}$) at the sidereal frequency. These motions generate periodic signals in experiments at the related frequencies with amplitudes proportional to the coefficients for Lorentz violation. Extracting these amplitudes from experimental data is the key measurement technique in the experiments that we consider here.

Most experiments to date have taken advantage of the rotation of the experiment at the sidereal frequency only, with some taking advantage of approximate boost effects considered at linear order in the boost velocity (see, e.g., Refs. 9–13) and even fewer considering effects at second order in boost effects.[14] Searches for anomalous spin-precession effects are among the most sensitive searches in the fermion sector. We have shown that utilizing boost effects beyond linear order in a reinterpretation of these highly sensitive tests results in complete coverage of the 44-degree-of-freedom coefficient space.[7,15]

In what follows, we use the standard SME convention in which lower-case indices denote components of tensors in the lab frame and upper-case indices denote components of tensors in the Sun-centered frame. To express $\tilde{b}_j$ in terms of the constant Sun-centered frame coefficients, we must use the transformation matrix $A_\mu{}^\Xi$ that takes us from the Sun-centered frame to the lab frame to transform each of the SME coefficients that make up $\tilde{b}_j$. Hence the following substitutions are used in Eq. (3):

$$b_j = A_j{}^\Xi b_\Xi \tag{4}$$

$$d_{jt} = A_j{}^\Xi A_t{}^\Sigma d_{\Xi\Sigma} \tag{5}$$

$$H_{kl} = A_k{}^\Xi A_l{}^\Sigma H_{\Xi\Sigma} \tag{6}$$

$$g_{klt} = A_k{}^\Xi A_l{}^\Sigma A_t{}^\Pi g_{\Xi\Sigma\Pi}. \tag{7}$$

The result can then be rewritten in terms of tilde coefficients in the Sun-centered frame and Fourier decomposed to show the frequencies at which time dependence of the experimental signals arises. The amplitudes of the Fourier components are then proportional to the coefficients associated with each of those signals. The full transformation matrix is composed as

$\mathbf{A} = \mathbf{R}\mathbf{\Lambda_L}\mathbf{\Lambda_\oplus}$, where

$$\mathbf{R} = \begin{pmatrix} 1 & 0 & 0 & 0 \\ 0 & \cos\chi\cos\omega t & \cos\chi\sin\omega t & -\sin\chi \\ 0 & -\sin\omega t & \cos\omega t & 0 \\ 0 & \sin\chi\cos\omega t & \sin\chi\sin\omega t & \cos\chi \end{pmatrix} \tag{8}$$

is the rotation that takes us to a lab-fixed coordinate system with the z-axis pointing vertically upward in the lab. The angle χ is the colatitude of the experiment, which we take as zero for simplicity in the examples presented in this work. The remainder of the transformation involves boost matrices $\mathbf{\Lambda_\oplus}$ and $\mathbf{\Lambda_L}$ of the form

$$\mathbf{\Lambda} = \begin{pmatrix} \gamma & \gamma\beta_1 & \gamma\beta_2 & \gamma\beta_3 \\ \gamma\beta_1 & 1+\frac{(\gamma-1)\beta_1^2}{\beta^2} & \frac{(\gamma-1)\beta_1\beta_2}{\beta^2} & \frac{(\gamma-1)\beta_1\beta_3}{\beta^2} \\ \gamma\beta_2 & \frac{(\gamma-1)\beta_1\beta_2}{\beta^2} & 1+\frac{(\gamma-1)\beta_2^2}{\beta^2} & \frac{(\gamma-1)\beta_2\beta_3}{\beta^2} \\ \gamma\beta_3 & \frac{(\gamma-1)\beta_1\beta_3}{\beta^2} & \frac{(\gamma-1)\beta_2\beta_3}{\beta^2} & 1+\frac{(\gamma-1)\beta_3^2}{\beta^2}. \end{pmatrix} \tag{9}$$

Here, $\beta_1, \beta_2, \beta_3$ are the components of the respective boost velocities $\vec{\beta}_\oplus$ and $\vec{\beta}_L$. In the examples to follow, we work with a circular orbit approximation for Earth's orbit around the Sun, where

$$\vec{\beta}_\oplus = \beta_\oplus(\sin\Omega t, -\cos\eta\cos\Omega t, -\sin\eta\cos\Omega t), \tag{10}$$

and η is the inclination of Earth's orbit in the Sun-centered frame. We also set aside consideration of β_L effects.

3. The Sidereal Frequency and New Discovery Potential

It is perhaps surprising that consideration of effects suppressed by the annual boost of the Earth's revolution around the Sun leads to additional signals at the sidereal frequency. However, this feature indeed occurs and results in new limits on Lorentz violation that can be read from existing published results. It may be similarly surprising that considering higher boost suppressions leads to added discovery potential in "old" experiments, but this also happens. In this section we illustrate how these effects arise using a simple limit of the full SME. We should emphasize that this material is a useful illustration of how more complicated results arise in the full SME and is not intended for experimental analysis, which should instead be pursued in the context of the full study.[7]

Suppose that d_{TX} and d_{XT} are the only nonzero Sun-centered frame coefficients. In this limit, the only nonzero tilde coefficients[a] in the Sun-centered frame are $\tilde{b}_X = -md_{XT}$ and $\tilde{d}_X = m(d_{TX} + \frac{1}{2}d_{XT})$. Consider contributions to $\tilde{b}_x$ up to order $\beta_\oplus^2$ arising from what remains of the calculation described in Eqs. (4)–(7)

$$\tilde{b}_x = -mA_x{}^X A_t{}^T d_{XT} - mA_x{}^T A_t{}^X d_{TX}. \tag{11}$$

To make further progress, we write the rotation and boost parts of this transformation explicitly. In order to do so, we must introduce the intermediate coordinate system, one that moves with the Earth but is nonrotating, denoted with a bar

$$\begin{aligned}\tilde{b}_x &= -mR_x{}^{\bar{x}}\Big[(\Lambda_\oplus)_{\bar{x}}{}^X(\Lambda_\oplus)_t{}^T d_{XT} + (\Lambda_\oplus)_{\bar{x}}{}^T(\Lambda_\oplus)_t{}^X d_{TX}\Big] \\ &\quad -mR_x{}^{\bar{y}}\Big[(\Lambda_\oplus)_{\bar{y}}{}^X(\Lambda_\oplus)_t{}^T d_{XT} + (\Lambda_\oplus)_{\bar{y}}{}^T(\Lambda_\oplus)_t{}^X d_{TX}\Big].\end{aligned} \tag{12}$$

Inserting the relevant matrix components yields

$$\begin{aligned}\tilde{b}_x &= -m\cos\omega t\Big[(1+\tfrac{1}{2}\beta_\oplus^2)d_{XT} + \beta_\oplus^2\sin^2\Omega t(\tfrac{1}{2}d_{XT} + d_{TX})\Big] \\ &\quad -m\sin\omega t\Big[-\beta_\oplus^2\cos\eta\sin\Omega t\cos\Omega t(\tfrac{1}{2}d_{XT} + d_{TX})\Big].\end{aligned} \tag{13}$$

We can express this result in terms of Sun-centered frame tilde coefficients as

$$\begin{aligned}\tilde{b}_x &= (1+\tfrac{1}{2}\beta_\oplus^2)\tilde{b}_X\cos\omega t \\ &\quad -\beta_\oplus^2\tilde{d}_X\cos\omega t\sin^2\Omega t + \beta_\oplus^2\tilde{d}_X\cos\eta\sin\omega t\sin\Omega t\cos\Omega t,\end{aligned} \tag{14}$$

which can be decomposed by Fourier component as

$$\begin{aligned}\tilde{b}_x = {} & \Big[(1+\tfrac{1}{2}\beta_\oplus^2)\tilde{b}_X - \tfrac{1}{2}\beta_\oplus^2\tilde{d}_X\Big]\cos\omega t + \tfrac{1}{4}\beta_\oplus^2\tilde{d}_X(1+\cos\eta)\cos(2\Omega t - \omega t) \\ & + \tfrac{1}{4}\beta_\oplus^2\tilde{d}_X(1-\cos\eta)\cos(2\Omega t + \omega t).\end{aligned} \tag{15}$$

Though our work here is in a comparatively simple limit of the full SME, the steps carried out above illustrate the same steps used in the full analysis.[7]

Because the SME is a test framework and not a model, one must choose how many coefficients to consider at one time in a fit to data. This is essential given that the full SME is an infinite series. One common choice that provides a sense of the basic level of depth to which a coefficient has been

[a]Note that in this limit $\tilde{b}_X^* = -\tilde{b}_X$ does not characterize an independent degree of freedom.

probed is a Maximum-Reach Analysis[13] in which each Lorentz-violating degree of freedom is fit to data separately under the assumption that all of the other coefficients are zero. Applying this approach in the context of Eq. (15) plays out as follows in the neutron sector. Prior analysis using the $\beta_\oplus = 0$ limit placed the constraint $\tilde{b}_X < 10^{-33}$ GeV by searching for the $\cos\omega t$ Fourier component,[4] while being insensitive to $\tilde{d}_X$. Here we see that in a Maximum-Reach Analysis, setting $\tilde{d}_X = 0$ and considering $\tilde{b}_x$ measurements generates the same result as existing studies from the $\cos\omega t$ Fourier component, while the reverse, setting $\tilde{b}_X = 0$ and constraining $\tilde{d}_X$ using the $\cos\omega t$ Fourier component, results in a limit of $\tilde{d}_X < 10^{-25}$ GeV. Hence we demonstrate that by considering additional powers of the boost velocity, a Maximum-Reach Analysis results in additional bounds on new coefficients using existing published results established by searching for sidereal variations.

Because the original experiments were so highly sensitive, the new bound we achieve after suffering a suppression of eight orders of magnitude is still arguably beyond the level one might expect for Planck-suppressed physics. For a coefficient with units of GeV as we have here, we might, as one possibility, generate a crude dimensional argument for the scale of Planck-suppressed physics as follows

$$\frac{m_n^2}{m_{\text{Planck}}} \sim 10^{-19}\,\text{GeV}, \tag{16}$$

where $m_n \sim 1$ GeV is the neutron mass and $m_{\text{Planck}} \sim 10^{19}$ GeV is the Planck mass.

Another common approach in the literature is to consider fitting some larger subset of SME coefficients to data together. This is known as a Coefficient-Separation Approach.[13] In the context of the current example, suppose we consider $\tilde{b}_X$ and $\tilde{d}_X$ together. This approach will require two independent measurements of different linear combinations of $\tilde{b}_X$ and $\tilde{d}_X$ to obtain independent measurements of both coefficients. In principle this can be done by searching for the $\cos\omega t$ Fourier component along with at least one of the $\cos(2\Omega t \pm \omega t)$ Fourier components. This illustrates that the methods we describe here open additional discovery potential for new analysis and/or experiments. Here one could, for example, imagine the possibility that $(1 + \frac{1}{2}\beta_\oplus^2)\tilde{b}_X = \frac{1}{2}\beta_\oplus^2\tilde{d}_X$. Thus Lorentz violation would be absent in searches at the sidereal frequency but observable at $2\Omega t \pm \omega t$. In the context of this example, some of these features may be difficult to observe because the frequencies are close together, requiring a large amount of data to separately observe the amplitudes. The point is to illustrate how some of the features seen in a more comprehensive approach[7,15] arise in the context of a tractable example.

4. The $g_{\lambda\mu\nu}$ Coefficients

There are many components of $g_{\Xi\Sigma\Pi}$ that have not yet been experimentally explored[3] that we succeed in accessing with higher-precision boost transformations.[7,15] It may be surprising that higher-precision boost transformations generate such access. This is a second feature of our work that is interesting to explore in the context of a simplified example. Let us take the example of $\tilde{g}_{TX} = m(g_{YTZ} + g_{ZTY})$, which is one such unexplored coefficient. A simple limit of the full SME in which this coefficient is the only nonzero tilde coefficient in the Sun-centered frame is to consider $g_{YTZ} = g_{ZTY} = -g_{TYZ} = -g_{TZY}$ as the only nonzero SME coefficients.

In this context, the nonzero contributions to $\tilde{b}_x$ can be written

$$\tilde{b}_x = m g_{yzt}, \tag{17}$$

or in terms of Sun-centered frame coefficients

$$\tilde{b}_x = m A_y{}^{\Xi} A_z{}^{\Sigma} A_t{}^{\Pi} g_{\Xi\Sigma\Pi} \tag{18}$$

$$\begin{aligned} &= m A_y{}^{Y} A_z{}^{T} A_t{}^{Z} g_{YTZ} + m A_y{}^{T} A_z{}^{Z} A_t{}^{Y} g_{TZY} \\ &\quad + m A_y{}^{Z} A_z{}^{T} A_t{}^{Y} g_{ZTY} + m A_y{}^{T} A_z{}^{Y} A_t{}^{Z} g_{TYZ}. \end{aligned} \tag{19}$$

Let us focus on observables that appear as the $\cos\omega t$ Fourier component of the signal. The $\cos\omega t$ Fourier components in Eq. (19) arise from the transformation matrices of the form $A_y{}^{\Xi}$ that appear in each term. This factor can be expanded as follows

$$A_y{}^{\Xi} = R_y{}^{\bar{y}} (\Lambda_\oplus)_{\bar{y}}{}^{\Xi} + R_y{}^{\bar{x}} (\Lambda_\oplus)_{\bar{x}}{}^{\Xi}. \tag{20}$$

We will focus here on the first term in Eq. (20), setting aside the second term which produces $\sin\omega t$. The rotation factors in the other $A_\mu{}^{\Xi}$ appearing in Eq. (19) are one.

Inspection of the second line of Eq. (19) reveals that all three of the boost matrix components are off diagonal, meaning that these terms generate contributions beyond quadratic order in $\beta_\oplus$ while the first line involves one on-diagonal and two off diagonal boost matrix components. Thus the following quadratic contributions to $\tilde{b}_x$ involving $\cos\omega t$ arise

$$\tilde{b}_x \subset m\cos\omega t\Big[(\beta_\oplus)^2_Z g_{YTZ} + (\beta_\oplus)^2_Y g_{TZY}\Big]. \tag{21}$$

Using the antisymmetry of $g_{\lambda\mu\nu}$ on the first two indices and the explicit form of the boost components yields

$$\tilde{b}_x \subset m\cos\omega t\beta^2_\oplus(\sin^2\eta g_{YTZ} - \cos^2\eta g_{ZTY})\cos^2\Omega t. \tag{22}$$

Using the defining equality of the special limit that we consider here, limiting attention to the $\cos\omega t$ Fourier component, and rewriting in terms of $\tilde{g}_{TX}$, we find

$$\tilde{b}_x \subset -\tfrac{1}{4}\cos\omega t\beta^2_\oplus\cos 2\eta\tilde{g}_{TX}, \tag{23}$$

revealing that indeed our sample unmeasured $g_{\Xi\Sigma\Pi}$-containing tilde coefficient, $\tilde{g}_{TX}$, does appear in $\tilde{b}_j$ searches, and in fact does so at the sidereal frequency at two boost suppressions.

5. Discussion

In this work we provide some examples of the kinds of access to coefficients that can be gained when considering higher-precision boost transformations. We show that new combinations of SME coefficients can be accessed at the commonly used sidereal frequency and that the search for variations at additional frequencies provides additional discovery potential. We have also demonstrated that components of the $g_{\lambda\mu\nu}$ coefficient that have never been observed arise in a higher-precision boost analysis in a relatively straightforward way. These examples are intended to provide an enhanced conceptual understanding of the results that appear in our more comprehensive work.[7,15]

There are other interesting features that emerge in a higher-precision boost analysis beyond those highlighted in the examples provided here. One notable case is the accessibility of the $\tilde{b}^*_J$ coefficient that is typically thought of as ripe for exploration with antimatter, as has been done for the proton.[16] Suppressed access to this coefficient by our methods complements existing approaches to exploring this coefficient with ordinary matter.[17] A special limit of the SME can be constructed in which $\tilde{b}^*_J$ is the only nonzero Lorentz-violating degree of freedom. This forms an interesting example for potential future work, though its complexity may limit its value as an illuminating example.

In this work we focus our examples on the simplified limit of a circular orbit for the Earth. Full results in the context of this model can be found in Ref. 15. In our comprehensive treatment, we consider a more complicated model of Earth's motion including effects such as the ellipticity of the Earth's orbit in the context of the full coefficient space for protons, neutrons, and electrons. We also consider implications for the muon sector in other work.[18,19] The consideration of higher-precision boost transformations is a surprisingly powerful and versatile tool in the ongoing search for Lorentz violation.

References

1. D. Colladay and V.A. Kostelecký, *Phys. Rev. D* **58**, 116002 (1998).
2. V.A. Kostelecký, *Phys. Rev. D* **69**, 105009 (2004).
3. V.A. Kostelecký and N. Russell, *Data Tables for Lorentz and CPT Violation,* 2025 edition, arXiv:0801.0287v18.
4. F. Allmendinger *et al.*, *Phys. Rev. Lett.* **112**, 110801 (2014).
5. J. Schaefer *et al.*, these proceedings; A. Ray *et al.*, these proceedings; A. Ray *et al.*, *Phys. Rev. D* **110**, 122001 (2024).
6. D. Amram *et al.*, *Phys. Rev. Lett.* **132**, 211801 (2024).
7. M.J. Basson *et al.*, in preparation.
8. B. Altschul, *J. Phys. A* **39**, 13757 (2006).
9. B.R. Heckel *et al.*, *Phys. Rev. D* **78**, 092006 (2008).
10. Y.V. Stadnik and V.V. Flambaum, *Eur. Phys. J. C* **75**, 110 (2015); J.M. Brown *et al.*, *Phys. Rev. Lett.* **105**, 151604 (2010); C. Gemmel *et al.*, *Phys. Rev. D* **82**, 111901(R) (2010); F. Canè *et al.*, *Phys. Rev. Lett.* **93**, 230801 (2004); S.K. Peck *et al.*, *Phys. Rev. A* **86**, 012109 (2012).
11. V.A. Kostelecký and C.D. Lane, *J. Math. Phys.* **40**, 6245 (1999).
12. A.J. Acevedo-Lopez, Y. Ding, K. Iwasaki, and A.J. Vargas, arXiv:2308.14201; C. Sanner *et al.*, *Nature* **567**, 204 (2019).
13. N.A. Flowers, C. Goodge, and J.D. Tasson, *Phys. Rev. Lett.* **119**, 201101 (2017).
14. H. Pihan-Le Bars *et al.*, *Phys. Rev. D* **95**, 075026 (2017); V.A. Kostelecký, A.C. Melissinos, and M. Mewes, *Phys. Lett. B* **761**, 1 (2016).
15. F. Lopez, Z. Zhang, B.R. Lott, and J.D. Tasson, these proceedings.
16. C. Smorra and A. Mooser, *J. Phys. Conf. Ser.* **1412**, 032001 (2020); C. Smorra *et al.*, BASE Collaboration, *Nature* **575**, 310 (2019).
17. A. Fittante and N. Russell, *J. Phys. G* **39**, 125004 (2012).
18. C. Holl *et al.*, in preparation.
19. C. Holl *et al.*, these proceedings.

Chapter 14

New Limits on CPT-Symmetry Violation in Charm Mesons

W. Krzemień,* M. Kmieć,† A. Szabelski,* and W. Wiślicki,†

**High Energy Physics Division, National Centre for Nuclear Research, Andrzeja Sołtana 7, Otwock, Świerk, PL-05-400, Poland*

†Department of Complex Systems, National Centre for Nuclear Research, Andrzeja Sołtana 7, Otwock, Świerk, PL-05-400, Poland

Neutral flavor meson oscillations are a fascinating quantum phenomenon that allows for high-precision tests of fundamental symmetries, including CPT invariance. In this contribution, we present the methodology for testing CPT violation in neutral flavor meson systems, with particular emphasis on the charm sector. We report recently established constraints on CPT-violating effects, derived from a reinterpretation of the LHCb measurement of the time-dependent asymmetry in the Cabibbo-favored decays $D^0 \to K^-\pi^+$ and $\bar{D}^0 \to K^+\pi^-$. Our results improve the previous bounds by the FOCUS experiment by approximately two orders of magnitude. We also discuss prospects for CPT-invariance tests, including within the Standard-Model Extension framework.

1. Introduction

Neutral flavor meson oscillations represent a remarkable quantum-interference phenomenon, offering a sensitive probe of fundamental symmetries such as CPT invariance. Among known systems, neutral mesons, especially kaons, enable experimental sensitivity approaching the Planck scale.[1]

The time evolution of neutral mesons is typically described using the Weisskopf–Wigner approximation,[2] where the dynamics is governed by a 2×2 non-Hermitian effective Hamiltonian. CPT violation can be parametrized via a complex parameter z. It can be expressed (see, e.g., Ref. 3) as the ratio of the difference of the diagonal elements of the effective Hamiltonian, encoding mass (δm) and decay-width ($\delta\Gamma$) differences between flavor states, to the normalized differences of the mass and decay-width eigenvalues of the meson mass states. Defining the mixing parameters $x = \Delta M/\Gamma$ and $y = \Delta\Gamma/(2\Gamma)$, with Γ the average decay rate, one has

$$z = \frac{\delta m - i\,\delta\Gamma/2}{\Gamma(x - iy)}. \tag{1}$$

The parameter z is nonzero if and only if CPT is violated; see Refs. 4, 5. The parameters x and y are often called mixing parameters because they govern the oscillation behavior of the neutral-meson system.

In the classical approach, the parameter z is treated phenomenologically, without an underlying theoretical interpretation. Alternatively, z can be derived within a framework such as the Standard-Model Extension (SME),[6,16,17] which connects it to fundamental CPT- and Lorentz-violating mechanisms. Although the same formalism describes all four neutral-meson families (K^0, D^0, B^0, B_s^0), their phenomenology varies significantly due to differences in oscillation frequencies and decay widths. For example, the B_s meson changes flavor approximately 26 times within its lifetime, while oscillations in the charm sector are extremely slow, with a frequency of order 0.1%.

Various experimental strategies exist to study CPT violation in these systems, depending on whether the mesons are produced in correlated pairs (originating from quarkonium decays, e.g., $\phi(1020) \to K^0\bar{K}^0$, $\psi(3770) \to D^0\bar{D}^0$, $\Upsilon(4S) \to B^0\bar{B}^0$) or as uncorrelated particles, and whether they decay into flavor-specific or common final states.[3,7–9]

2. CPT in the Charm Sector

The charm sector offers a distinct and complementary environment for CPT-violation studies compared to the strange and beauty sectors. The mass of the charm quark (≈ 1.27 GeV/c^2) places it in the intermediate regime between light (u, d, s) and heavy (b, t) quarks, where nonperturbative QCD effects play a significant role. This fact, combined with various suppression mechanisms, makes theoretical predictions in charm particularly challenging.[10] In the Standard Model, both quantum oscillations and CP-violating effects in the charm sector are expected to be highly suppressed, typically at the level 10^{-3}–10^{-4}; see Ref. 11.

From an experimental point of view, several charm-specific features complicate CPT-violation studies. Due to the short lifetime of D^0 mesons ($\tau_{D^0} = 410$ fs) and slow oscillation rates, experiments must collect large datasets while maintaining high signal purity. This requires efficient and selective trigger systems, precise vertex reconstruction to distinguish prompt mesons from secondary ones (originating from B decays), excellent momentum resolution, and robust particle identification. Time-dependent analyses additionally benefit from large Lorentz boosts and detectors with ultrahigh time resolution. These conditions are exceptionally well fulfilled by the LHCb experiment, which provides large charm datasets and decay-time resolutions at the $0.1\,\tau_{D^0}$ level.[12]

A CPT-violation analysis can be carried out based on the two-hadron decays of D^0 mesons: $D^0 \to K^-\pi^+$ and its CPT-conjugate $\bar{D}^0 \to K^+\pi^-$. The decay $D^0 \to K^-\pi^+$ is dominated by the direct Cabibbo-favored (CF) amplitude, making it almost fully flavor-specific and thus a clean probe of CPT symmetry. The only contamination arises when D^0 oscillates into $\bar{D}^0$ and subsequently decays via the doubly Cabibbo-suppressed (DCS) mode $\bar{D}^0 \to K^-\pi^+$. However, the smallness of the DCS branching fraction, 1.36×10^{-4} compared to 3.947×10^{-2} for CF,[1] combined with slow mixing (the oscillation period is three orders of magnitude larger than the D^0 lifetime) makes this admixture typically negligible.

The central idea of the CPT test is the construction of a time-dependent decay-rate asymmetry,

$$A_{RS}(t) = \frac{N(D^0 \to K^-\pi^+)(t) - N(\bar{D}^0 \to K^+\pi^-)(t)}{N(D^0 \to K^-\pi^+)(t) + N(\bar{D}^0 \to K^+\pi^-)(t)}. \tag{2}$$

The CPT-violating parameter z is extracted by fitting this observable to the expected time dependence, which for flavor-specific channels can be written as[3]

$$A_{\mathrm{CPT}}(t) = A_{\mathrm{dir}} + \frac{2\,\Re(z)\,\sinh(\Delta\Gamma\, t/2) - 2\,\Im(z)\,\sin(\Delta m\, t)}{(1+|z|^2)\cosh(\Delta\Gamma\, t/2) + (1-|z|^2)\cos(\Delta m\, t)}, \tag{3}$$

where all asymmetries, including the direct CPT-violating term A_{dir}, are assumed to be much smaller than one. In this limit, the CPT-violation measurement is independent of CP-violation effects.

Due to the slowness of the $D^0\bar{D}^0$ oscillations, Eq. (3) can be expanded to leading order for $|x|\,\Gamma t \ll 1$ and $|y|\,\Gamma t \ll 1$, yielding

$$A_{\mathrm{CPT}}(t) = A_{\mathrm{dir}} + (\Re(z)\, y - \Im(z)\, x)\,\Gamma t. \tag{4}$$

3. New Classical Limits with Charm Mesons

Recently, new limits on CPT violation in the charm sector have been established.[4] The bounds were derived from a time-dependent asymmetry analysis of the $D^0 \to K^-\pi^+$ decay channel based on the data set collected by the LHCb collaboration.[13] The slope of the linear fit in Eq. (4) can be identified as $y\,\Re(z) - x\,\Im(z)$. The result is statistically consistent with zero, indicating no evidence of CPT violation at the current sensitivity level.

The measured slope s defines a linear relationship between $\Re(z)$ and $\Im(z)$ in the $(\Re(z), \Im(z))$ plane,

$$\Im(z) = \frac{y}{x}\Re(z) - \frac{s}{x},$$

as shown in Fig. 1 (left). This constraint can be translated into limits on the mass and decay-width differences between flavor and antiflavor states (Fig. 1, right). The resulting limits lie in the range 10^{-15}–10^{-16} GeV, representing the strongest classical constraint on CPT violation in the charm sector to date.

At the present level of experimental precision, the dominant uncertainty arises from limited knowledge of potential CP-violating effects. While their current impact remains subdominant, they could become significant in future analyses with increased statistics.

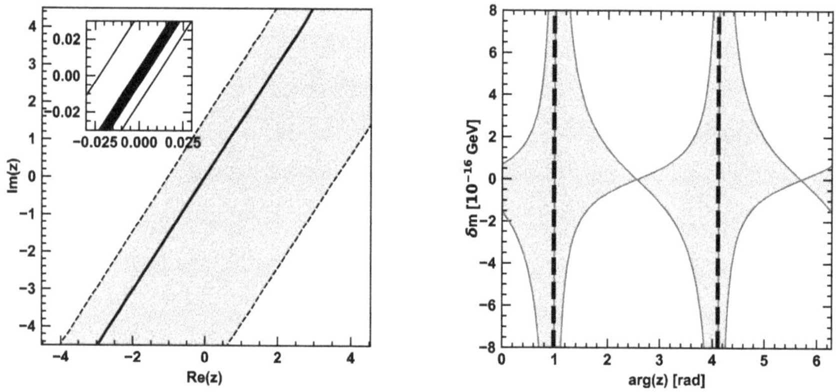

Fig. 1. Experimental limits on CPT violation in charm. Left: constraints on the parameter z at 95% CL in the $D^0 \to K^-\pi^+$ channel from FOCUS[14] (gray) and inferred from LHCb[13] (white). The uncertainty from possible Standard-Model CP violation is shown in black. Right: mass difference δm versus $\phi = \arg(z)$; the gray region corresponds to 95% CL, and the dashed lines indicate values where δm is undefined due to the vanishing of the linear asymmetry term.
Source: Adapted from Ref. 4.

4. Prospects for CPT Tests in Charm Within the SME

The SME[6,16–18] is an established framework for CPT-violation searches across diverse physical systems. It incorporates the idea, due to Greenberg,[15] that any CPT violation implies Lorentz-symmetry breaking. In the SME, CPT- and Lorentz-violating terms appear as perturbative corrections to the Standard Model and General Relativity, preserving the core properties of effective quantum field theory and enabling systematic experimental tests.

In the SME picture, Lorentz violation manifests through background tensor fields permeating spacetime to which particles can couple. This breaks isotropy and can introduce dependencies on the particle boost and momentum direction. A key consequence is the prediction of sidereal variations arising from Earth's rotation, which changes the experiment's orientation relative to the background field. These characteristic dependencies can both enhance sensitivity and provide distinctive signatures.

In particular, the CPT-violating parameter z is not treated as a constant but depends on the meson boost and momentum four-direction β^μ and can be expressed as[17]

$$z = \frac{\Delta a_\mu \beta^\mu}{\Gamma(x - iy)}, \tag{5}$$

where Δa_μ are real SME coefficients.

The only experimental limits on CPT violation in the charm sector within the SME framework to date were set by the FOCUS collaboration.[14] FOCUS analyzed $D^0 \to K^-\pi^+$ using data collected at Fermilab from a high-energy photon beam with energy ≈ 180 GeV incident on a fixed BeO target. The study was performed using both classical and SME-based approaches. The average Lorentz boost of the D^0 mesons in the sample was approximately 39, and the statistics were modest, with about 35k events.

Between 2011 and 2018, the LHCb experiment collected $\sim 10^4$ times more events in the same decay mode. This dramatic increase in statistics, coupled with similar boosts, suggests a potential improvement by roughly a factor of 100 in sensitivity to the SME coefficients Δa_μ for this channel.

However, in the SME framework the leading linear term $(y\,\Re z - x\,\Im z)$ can vanish under the assumption that Δa_μ are real, and a more complete treatment is required. Higher-order terms must be included, previously neglected DCS contributions become relevant, and CP-violation effects must be incorporated. An analysis of the LHCb dataset using the full SME-based model is currently ongoing.

5. Conclusions and Future Perspective

Neutral flavor meson oscillations enable high-precision tests of fundamental symmetries, including CPT invariance. In the kaon system, experimental sensitivities approach the Planck scale.

We have recently established new classical limits on CPT violation in the charm sector. Our results improve the previous FOCUS bounds by about two orders of magnitude and translate to upper limits on the mass and decay-width differences between flavor particle and antiparticle states at the level 10^{-15}–10^{-16} GeV at 95% CL.

These results complement CPT-violation studies in other meson systems.[1] The attainable precision on δm and $\delta\Gamma$ depends on the interplay between the lifetime scale and the mixing parameters, which differ considerably between meson families. For instance, in the neutral-kaon system the small denominator in Eq. (1) strongly amplifies sensitivity to δm and $\delta\Gamma$, reaching about 10^{-18} GeV,[19] while in the beauty sector current bounds range from $\sim 10^{-14}$ GeV for B[20,21] to $\sim 10^{-12}$ GeV for B_s.[22] Since CPT-violating couplings could depend on quark content, it remains valuable to explore all flavor sectors.[3]

An ongoing analysis of LHCb data within the SME framework is further exploring CPT-violating effects in charm. In addition, recent theoretical progress has made data analyses within the nonminimal SME approach feasible.[23]

In the beauty sector, the data already recorded by LHCb should improve previous SME-based CPT-violation limits by at least a factor of two, with even greater potential as data taking progresses. In the charm sector, a principal limitation remains the imprecisely known contribution from possible CP violation; this may be measured more precisely in forthcoming experiments.

The BESIII experiment[24,25] operates at the $\psi(3770)$ resonance and has recorded large samples of quantum-correlated $D^0\bar{D}^0$ pairs. Since the resonance is produced essentially at rest, analyses relying on large boosts are not feasible. The planned Super Tau–Charm Facility[26] is expected to significantly enhance sensitivity by combining high-statistics correlated charm production with boosted kinematics, enabling interferometric studies well suited for CPT studies. KLOE/KLOE-2,[27,28] which has recorded the world's largest sample of entangled kaon pairs, is currently analyzing its full dataset[29,30] and may contribute further to CPT searches in kaons. Belle II[31,32] is now taking data and may improve selected CPT limits in charm and beauty, especially in channels with neutral final-state particles.

These datasets provide a rich and promising landscape to further tighten constraints and potentially discover CPT-symmetry violation.

Acknowledgments

W.K. wishes to thank the organizers for the kind invitation to present a talk and for stimulating discussions during the meeting.

References

1. R.L. Workman *et al.* (Particle Data Group), *Prog. Theor. Exp. Phys.* **2022**, 083C01 (2022).
2. V. Weisskopf and E. Wigner, *Z. Phys.* **63**, 54 (1930).
3. V.A. Kostelecký, *Phys. Rev. D* **64**, 076001 (2001).
4. W. Krzemień, M. Kmieć, A. Szabelski, and W. Wiślicki, *Phys. Rev. D* **110**, 055021 (2024).
5. Y. Amhis *et al.*, *Phys. Rev. D* **107**, 052008 (2023).
6. D. Colladay and V.A. Kostelecký, *Phys. Rev. D* **55**, 6760 (1997).
7. J. Bernabeu, N.E. Mavromatos, and J. Papavassiliou, *Phys. Rev. Lett.* **92**, 131601 (2004).
8. Á. Roberts, *Phys. Rev. D* **96**, 116015 (2017).
9. J. Van Tilburg and M. Van Veghel, *Phys. Lett. B* **742**, 236 (2015).
10. D. Friday, E. Gersabeck, A. Lenz, and M.L. Piscopo, *Charm Physics*, arXiv:2506.15584.
11. A. Lenz and G. Wilkinson, *Annu. Rev. Nucl. Part. Sci.* **71**, 59 (2021).
12. T. Pajero, *Mod. Phys. Lett. A* **37**, 2230012 (2022).
13. R. Aaij *et al.* (LHCb Collaboration), *Phys. Rev. D* **104**, 072010 (2021).
14. J.M. Link *et al.*, *Phys. Lett. B* **556**, 7 (2003).
15. O.W. Greenberg, *Phys. Rev. Lett.* **89**, 231602 (2002).
16. V.A. Kostelecký, *Phys. Rev. Lett.* **80**, 1818 (1998).
17. V.A. Kostelecký, *Phys. Rev. D* **61**, 016002 (1999).
18. V.A. Kostelecký and N. Russell, *Data Tables for Lorentz and CPT Violation*, *Rev. Mod. Phys.* **83**, 11 (2011); 2025 edition, arXiv:0801.0287v18.
19. A. Angelopoulos *et al.*, *Phys. Lett. B* **471**, 332 (1999).
20. T. Higuchi *et al.*, *Phys. Rev. D* **85**, 071105 (2012).
21. J.P. Lees *et al.*, *Phys. Rev. D* **94**, 011101 (2016).
22. R. Aaij *et al.* (LHCb Collaboration), *Phys. Rev. Lett.* **116**, 241601 (2016).
23. B.R. Edwards and V.A. Kostelecký, *Phys. Lett. B* **795**, 620 (2019).
24. BESIII public webpage, http://bes3.ihep.ac.cn.
25. M. Ablikim *et al.* (BESIII Collaboration), *Nucl. Instrum. Meth. A* **614**, 345 (2010).
26. M. Achasov *et al.*, *Front. Phys. (Beijing)* **19**, 14701 (2024).

27. KLOE-2 public website, https://w3.lnf.infn.it/research/particle-physics/kloe-2/?lang=en.
28. G. Amelino-Camelia *et al.* (KLOE-2 Collaboration), *Eur. Phys. J. C* **68**, 619 (2010).
29. A. Di Domenico, *The "Back from the Future" Quantum Effect in Entangled Neutral Kaons at KLOE/KLOE-2*, in *KAON 2025* (2025).
30. E. Czerwinski, *Updates on CP Symmetry Tests at KLOE-2*, in *KAON 2025* (2025).
31. Belle II public webpage, https://www.belle2.org.
32. T. Abe *et al.* (Belle II Collaboration), *Belle II Technical Design Report* (2010).

Chapter 15

Lorentz Violation and the Equivalence Problem of Finsler Geometry

Benjamin R. Edwards

Indiana University Center for Spacetime Symmetries,
Bloomington, IN 47405, USA

The link between Finsler geometry and physical systems with Lagrangians exhibiting apparent Lorentz violation has been well established. The Standard Model Extension, a theoretical framework extending standard physics to include tiny violations of Lorentz symmetry, has produced multiple ties to new and existing Finsler spaces already. The problem of characterizing Finsler spaces has been solved for comparatively simple Finsler spaces, but for many of the new SME-based Finsler spaces, this problem has remained an open question. The fact that SME backgrounds give rise to distinct physics strongly suggests that their related Finsler geometries can be distinguished with appropriate geometric quantities.

1. Introduction

Riemann–Finsler geometry[1,2] has been linked to effective field theories exhibiting Lorentz violation in multiple contexts within the Standard Model Extension (SME) framework.[3–7] The SME is a realistic effective field theory incorporating Lorentz and CPT violation with standard physics. Physical intuition developed in the context of the various limits of the SME may be exploited to assist with resolving the open geometric questions related to Finsler geometry. Finsler spaces can be created by mapping the classical path of a wave packet experiencing a Lorentz-violating background in a limit of the SME to a geodesic in a Riemann–Finsler space via continuation in a process similar to a Wick rotation. This correspondence can be exploited to produce physically relevant Finsler spaces in a way that takes advantage of the systematic classification of Lorentz-violating effects used within the SME, while simultaneously ensuring a physical context for the resulting spaces. One of the main problems of Finsler geometry has been the equivalence problem, i.e., the challenge of defining a complete set of invariants or being able to distinguish when two given Finsler metrics simply differ by a change of local coordinates. While this problem has been solved for some Finsler spaces, the lion's share of SME-related Finsler spaces already created by the process mentioned above have yet to be characterized.

2. SME-Based Finsler Spaces

Perturbative methods for deriving Finsler spaces that correspond to limits of the SME have been expanded over the years to include even the nonminimal sector, and were first described in Ref. 7 for the scalar effective field theory with Lagrange density

$$\mathcal{L}(\phi,\phi^\dagger) = \partial^\mu \phi^\dagger \partial_\mu \phi - m^2 \phi^\dagger \phi + \frac{1}{2}[\partial_\mu \phi^\dagger (\hat{k_c})^{\mu\nu} \partial_\nu \phi - i\phi^\dagger (\hat{k_a})^\mu \partial_\mu \phi + \text{h.c.}]. \tag{1}$$

This correspondence created a large class of SME-related spaces called k-spaces that generalized the already popular (α, β) metrics to include more complex direction dependence. The correspondence is made by recognizing the classical path of a wave packet propagating in a Lorentz-violating background that can then be mapped to the extremal action path of a point-particle Lagrangian. First, the spatial velocity of the particle is identified with the instantaneous group velocity of the wave packet in the field theory via $-u^j/u^0 = \partial p_0/\partial p_j$. Next, the Lagrangian is required to produce the analogous dispersion relation as the appropriate limit of the SME Lagrange density. Finally, homogeneity of degree one in the velocity is also required, so as to preserve path-length under reparameterization (i.e., $L = u^i p_i$). These conditions allow the momenta, and thus the Lagrangian itself, to be expressed in terms of the velocity only.

The resulting Lagrangian generally fails to meet all the criteria of a Finsler norm.[8] For one, the norms of vectors in spacetime can fail to be nonnegative and the resulting metric obtained by taking the Hessian of L^2

on the fibers is not positive definite. For another, the set of null vectors $S = \{y|F(y) = 0\}$ is nontrivial. The resulting space is instead a Lorentz–Finsler space, although various approaches to Lorentz–Finsler spaces have resulted in competing and nonequivalent definitions. Even in the case of Riemann–Finsler geometry, the restriction of smoothness, as well as the positive definiteness of the Finsler metric, for all $y \in TM \setminus S$ is too strict for the SME-related spaces. For Riemann–Finsler geometries corresponding to Lorentz violation, the presence of a Lorentz-violating background field naturally creates a preferred tensor background against which there will generically arise a subspace that fails to satisfy the typical Finsler requirements of smoothness and nonnegativity outside of the usual slit S. Nevertheless, a relaxing of the definition to stipulate smoothness and nonnegativity almost everywhere in each T_xM may be sufficient to include the examples of SME-related spaces presently being considered.

For the scalar field theories represented by Eq. (1), a large class of relatively unexplored Finsler spaces can be created. For example, the Finsler norm

$$F^{(d)} = y - \frac{1}{2} y\tilde{k}^{(d)} \tag{2}$$

represents the first-order Finsler space associated with the mass-dimension d coefficient for Lorentz violation in an n-dimensional space, where the norm of the tangent vector is y, and the $\tilde{k}^{(d)}$ is the background coefficient contracted with $d - n + 2$ velocity components.

Another large class of SME-related spaces are the so-called bipartite spaces. This important class of spaces arises in many limits of the SME. They are characterized by the equations

$$\begin{aligned} F &= \rho + \sigma \\ \rho &:= \sqrt{y^j r_{jk} y^k} \\ \sigma &:= \pm\sqrt{y^j s_{jk} y^k}, \end{aligned} \tag{3}$$

where r_{jk} is the Riemannian metric, and s_{jk} is a symmetric two-form. One such space is the special case where s_{jk} takes the form

$$s_{jk} = b^2 r_{jk} - b_j b_k, \tag{4}$$

giving rise to the SME-related b-space.

3. Finsler Geometry

The general line element in Finsler geometry can be written

$$ds = F(x^1, x^2, \ldots, x^n; dx^1, dx^2, \ldots, dx^n), \tag{5}$$

where F is a smooth nonnegative function on the slit tangent bundle $TM \setminus S$ that obeys the homogeneity requirement $F(x, \lambda y) = \lambda F(x, y)$. Although Riemann himself originally introduced this line element in 1854, the name Riemannian geometry has been assigned to the special case where $F^2 = g_{ij} dx^i dx^j$. It was not until Paul Finsler's 1928 thesis considered the more general case that Riemann originally introduced.[1,2]

Several objects of interest can be defined that are important for studying the space and its properties. First, we denote local coordinates on a manifold M with x^i, and tangent vectors in $T_x M$ with y^i. This allows local points in TM, viewed as a manifold, to be written (x^i, y^i). The first fundamental form can be calculated from the components of the metric, which are given by the y-Hessian $g_{ij} = (\frac{1}{2}F^2)_{y^i y^j}$. The homogeneity requirement on F forces all of the g_{ij} to be homogeneous of degree zero in y. With the inner product defined by g_{ij}, the pulled-back $T^\star M$ admits a globally defined zero-homogeneous geometric quantity of unit length

$$p = \frac{\partial F}{\partial y^i} dx^i \tag{6}$$

called the Hilbert form. In physics applications, where F plays the role of the Lagrangian for a point particle, the Hilbert form corresponds exactly to the momentum.

Because of the homogeneity conditions, the Hilbert form controls all aspects of the geometric structure. First, we may write $F = y^j p_j$, which allows one to identify the indicatrix $\mathcal{I} = \{y | F(y) = 1\}$. The indicatrix, in turn, is a central object in the study of a Finsler space since the homogeneity conditions dictate that all the geometric objects (that is to say, the large list of zero-homogeneous tensors on TTM, i.e., g_{ij}, $h_{ij} = FF_{jk}$, $C_{ijk} = \frac{1}{4}(F^2)_{jkl}$, $I_j = g^{jk} C_{jkl}$, p_j, $\ldots$) are scale invariant in $y \in T_x M$ at each x, and need only be defined on $\mathcal{I}$ to be defined in the entire tangent space. In this way, once one determines the Hilbert form, the entire remaining Finsler structure is fixed and fully determined.

4. Characterization of Finsler Spaces

The equivalence problem in local Finsler geometry is central to the study of Finsler spaces. When presented with two Finsler norms, the question naturally arises whether they represent distinct Finsler spaces or if they are simply the same space expressed in a different set of local coordinates. For Finsler geometry, all geometric quantities constructed from F are scale invariant (i.e., zero-homogeneous) in y. One such quantity, the Cartan form $C_{jkl} = \frac{1}{4}(F^2)_{jkl}$, serves the purpose of characterizing Riemannian spaces. For Riemannian spaces, $g = g(x)$ only and we have that $C = 0$. Another quantity that serves the same purpose is the Matsumoto torsion, $M_{jkl} = C_{jkl} - \frac{1}{n+1}\sum_{(jkl)} I_j h_{kl}$, which vanishes for Randers space, where the Riemannian norm ρ is augmented by a background one-form a and has the norm $F = \rho - a_j y^j$.

The above results extend to almost-Finsler spaces and the SME-related a space, a special case of bipartite space where $s_{jk} = a_j a_k$. Although similar to Randers, they differ in that a spaces remain unchanged under $y \to -y$ while Randers spaces do not. Nevertheless, the result that $M = 0$ for a spaces carries over under the almost-Finsler extension. Other bipartite spaces have been shown to produce nonvanishing Cartan and Matsumoto torsions, and until now have eluded the same resolution to the equivalence problem. Efforts to resolve this problem for more SME-related Finsler spaces are underway.[9]

References

1. B. Riemann, *Über die Hypothesen welche der Geometrie zu Grunde liegen*, in R. Baker, C. Christensen, and H. Orde, *Bernhard Riemann, Collected Papers*, Kendrick Press, Heber City, Utah, 2004.
2. P. Finsler, *Über Kurven und Flächen in allgemeinen Räumen*, University of Göttingen dissertation, 1918; Verlag Birkhäuser, Basel, Switzerland, 1951.
3. V.A. Kostelecký, *Phys. Lett. B* **701**, 137 (2011).
4. V.A. Kostelecký, N. Russell, and R. Tso, *Phys. Lett. B* **716**, 470 (2012).
5. V.A. Kostelecký and N. Russell, *Phys. Lett. B* **693**, 443 (2010).
6. J.A.A.S. Reis and M. Schreck, *Phys. Rev. D* **97**, 065019 (2018).
7. B.R. Edwards and V.A. Kostelecký, *Phys. Lett. B* **786**, 319 (2018).
8. S.S. Chern and Z. Shen, *Riemann-Finsler Geometry*, World Scientific, Singapore, 2005.
9. J. Davis, B.R. Edwards, and V.A. Kostelecký, arXiv:2508.21744.

Chapter 16

You Wouldn't Expect C, P, and T Operators to be Renormalized

B. Altschul

Department of Physics and Astronomy, University of South Carolina, Columbia, SC 29208, USA

There are redundancies in the effective field theory description of the SME. A fermion f term is actually equivalent to a certain type of c term, and this exact equivalence has a number of peculiar consequences. Although an f term appears superficially to be CPT odd, the f theory actually has no physical CPT violation, because the true C, P, and T operators are not the usual ones and instead depend explicitly on f. The renormalization group β-functions are also ambiguous and depend on the renormalization scheme, even at one-loop order. Combining these two unexpected observations produces a third one—that the C, P, and T operators may have nontrivial renormalization group scale dependences of their own.

1. Introduction

Special relativity underlies both of the theories that we use to describe fundamental physics: the Standard Model of particle physics and the general theory of relativity. The Lorentz symmetry of special relativity manifests itself in somewhat different ways in these two regimes, but it is a thread connecting both of them. Given the importance of this symmetry, there has long been significant interest in testing whether it holds exactly or merely approximately. This interest is quite natural. We already know that an apparent but, in reality, only approximate symmetry can have profound implications and provide key insights about where new fundamental physics might be found.

In this millennium, thanks primarily to advances in effective field theory methods, it has been possible to parameterize violations of rotation invariance and Lorentz boost invariance in a systematic way. This, in turn, has prompted experimental interest in many types of Lorentz violation that had previously flown under scientists' radar. These kinds of renewed experimental searches have not found any compelling evidence that Lorentz symmetry is not exact, but they remain ongoing.

The local effective field theory used for describing Lorentz violation with known Standard Model fields is prosaically known as the Standard-Model Extension (SME).[1,2] Because of the close connection between Lorentz invariance and CPT invariance,[3] the SME is also capable of describing generically any forms of CPT violation that are stable, unitary, local, and constructed out of the Standard Model's fermion and boson fields. (Gravity is slightly trickier.) The SME (or at least the particle theory part) is a quantum field theory (QFT), and properly understanding a QFT always entails understanding the characteristics of its $\mathcal{O}(\hbar)$ and higher loop corrections. The minimal SME constitutes the sub-theory of the full SME framework that should be expected to be renormalizable, as it contains only the finite number of local, Hermitian, and gauge-invariant operators that can be constructed to mass dimension no greater than four. The minimal SME action looks much like the ordinary Standard Model action, except that the field operators it contains are not constrained to have all their Lorentz indices fully contracted. This minimal theory is often the most logical framework for analyzing the results of experimental Lorentz and CPT tests. There has been a fair amount of good work on the elucidation of the renormalization properties of the minimal SME, especially at one-loop order.[4–12] However, our understanding of the theory at the quantum level is by no means complete.

2. Fermion Lorentz Violation

The free fermion sector of the SME contains the operators that control the propagation of spin-$\frac{1}{2}$ particles. In the minimal SME, with only superficially renormalizable operators, the Lagrange density for a single species of fermion is

$$\mathcal{L} = \bar{\psi}(i\Gamma^{\mu}\partial_{\mu} - M)\psi. \tag{1}$$

The Γ and M terms generalize the usual Dirac Lagrange density's γ and m, by including all possible matrix structures, contracted with the fixed vector and tensor backgrounds that describe the Lorentz violation. These background structures enter in the forms

$$\Gamma^\mu = \gamma^\mu + \Gamma_1^\mu = \gamma^\mu + c^{\nu\mu}\gamma_\nu + d^{\nu\mu}\gamma_5\gamma_\nu + e^\mu + if^\mu\gamma_5 + \tfrac{1}{2}g^{\lambda\nu\mu}\sigma_{\lambda\nu} \quad (2)$$

$$M = m + m'\gamma_5 + M_1 = m + m'\gamma_5 + a^\nu\gamma_\nu + b^\nu\gamma_5\gamma_\nu + \tfrac{1}{2}H^{\mu\nu}\sigma_{\mu\nu}. \quad (3)$$

Some of the coefficients in $\mathcal{L}$ are more interesting than others. For example, m' is not Lorentz violating, but rather represents a Majorana mass term that may be absorbed into m in the noninteracting Lorentz-invariant theory by a field redefinition.

Without the condition of Lorentz symmetry—which amounts to a condition that the operators in the Lagrange density of the theory should have no free Lorentz indices—there are many more dimension-three and dimension-four operators than in a conventional QFT. However, there is actually some redundancy in the SME description, and there exists an exact equivalence between two different types of Lorentz-violating fermion theories. This redundancy is at the root of all the observations we shall make in this paper. There is a transformation of the SME Lagrange density with solely an f term into one with just a c term instead.[13] The theory with f, defined by

$$\mathcal{L}_f = \bar{\psi}[i(\gamma^\mu + if^\mu\gamma_5)\partial_\mu - M]\psi \quad (4)$$

is converted into

$$\mathcal{L}_c = \bar{\psi}'[i(\gamma^\mu + c_{\text{eff}}^{\nu\mu}\gamma_\nu)\partial_\mu - M]\psi' \quad (5)$$

by a local linear transformation acting on the four-component fermion field,

$$\psi' = e^{\frac{i}{2}f^\nu\gamma_\nu\gamma_5 G(-f^2)}\psi, \quad (6)$$

where $G(\xi) = \frac{1}{\sqrt{\xi}}\tan^{-1}\sqrt{\xi}$. The c tensor in the transformed Lagrange density is

$$c_{\text{eff}}^{\nu\mu} = \frac{f^\mu f^\nu}{f^2}\left(\sqrt{1-f^2} - 1\right) \approx -\frac{1}{2}f^\nu f^\mu, \quad (7)$$

with the approximate form applying when $|f^2|$ is small.

The equivalence between f and c_{eff} has a number of interesting consequences—some relatively straightforward, others rather more subtle. While calculations may be performed in either version of the theory, in some cases it may be significantly easier to use one than the other. It seems that the c-type theory is typically simpler to work with, one reason for this being that effects that appear at first order in c_{eff} are necessarily second order in f. In fact, from the form of c_{eff}, it is clear that observable phenomena can only depend on even powers of the f coefficients, and this leads to another curious observation—that the individual components of the f operator do not have physically realizable CPT characteristics. While the operators appear, according to usual properties of Dirac bilinears, to be odd under CPT, there is no physical CPT violation, because odd powers of f never appear in observables.

Moreover, because the theories with f and c_{eff} are entirely equivalent, it seems that by understanding the quantum corrections in one version of the theory, we should be able to draw conclusions about the $\mathcal{O}(\hbar)$ behavior of the other. Taking the c and f terms to be physically equivalent, it would appear that this equivalence should manifest itself in the RG scalings of the two operator types. We shall express the RG β-function for a quantity x using $\beta_x = x\Psi(x)$. With this notation, it seems that at leading order, we ought to have $\Psi(f^\mu) = \frac{1}{2}\Psi(c^{\nu\rho})$, because this would make the RG evolution of $c^{\nu\mu}$ and $f^\mu f^\nu$ the same. However, explicit calculations of β-functions did not initially bear this out. Instead, it was found that (regardless of what type of interactions between fermions were introduced, gauge or Yukawa) β_f seemed to vanish at leading order, while β_c did not.

The reason for this discrepancy was initially somewhat tricky to track down.[14] It seemed clear that at least part of the difficulty had to be related to the fact that the explicit calculations of β_f were only performed to first order in the Lorentz violation coefficient, while the equivalence to the c theory appeared at second order. However, it is not necessarily obvious how this could make a difference in the concrete determination of a β-function. The ultimate reason this issue arises turns out to be that there are different ways that the counterterms may be chosen to cancel the divergences in the theory with nonzero f coefficients. A one-loop fermion self-energy diagram (Fig. 1) with two f insertions on the internal fermion line is, as naively expected, divergent. The divergence can be canceled in either of two

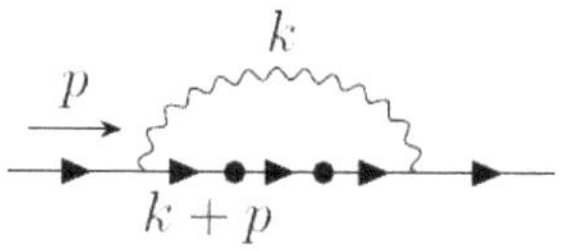

Fig. 1. Divergent one-loop diagram with two Lorentz-violating f vertex insertions on the internal fermion line.

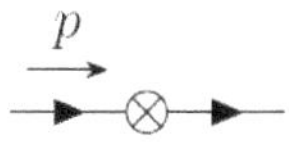

Fig. 2. Diagram with a single δc fermion insertion to cancel the $\mathcal{O}(f^2)$ divergence in the diagram from Fig. 1.

Fig. 3. The two counterterm diagrams that together cancel the divergence with alternate f and δf insertions.

different ways: by a single δc counterterm, of the type shown in Fig. 2, or by the pair of diagrams shown in Fig. 3, each with two insertions—one of f and one of δf. Since the prefactors of the divergent logarithms appearing in the counterterms determine the one-loop β-functions for the associated operators, this means that there are two different ways for the SME coefficients to evolve under changes of the renormalization scale.

In fact, there are more than just two choices for how to renormalize the theory. Linear combinations of the δc diagram and the paired δf diagrams may be used to eliminate the divergences. This means that there are actually a continuous family of one-loop renormalization schemes, with corresponding continuous families of β-function pairs. In a theory in which the force carriers exchanged between fermions are Yukawa bosons, the divergence in Fig. 1 may be canceled by any combination of counterterms

$$\delta_{c^{\nu\mu}} = -\frac{Xg^2}{6}\eta f^{\nu} f^{\mu} \tag{8}$$

$$\delta_{f^{\mu}} = -\frac{(1+2X)g^2}{6}\eta f^{\mu}, \tag{9}$$

where g is the Yukawa coupling, and

$$\eta = \frac{\Gamma(\epsilon/2)}{(4\pi)^{2-\epsilon/2}\mathcal{M}^{\epsilon}} \tag{10}$$

is the logarithmic divergence. Any value of the parameter X will work. $X = 0$ corresponds to the choice discussed previously, motivated by the observation that the f is physically equivalent to a c.

Note that if δc is nonzero, the theory must actually contain a tree-level c coefficient as well as an f, and there must be an additional contribution to (8) corresponding to the self-renormalization of the c operator. Including the contributions from both types of SME terms, the RG β-functions depend on the scheme parameter X according to

$$\beta_{c^{\nu\mu}} = \frac{2g^2}{3(4\pi)^2}c^{\nu\mu} - \frac{Xg^2}{3(4\pi)^2}f^\nu f^\mu \tag{11}$$

$$\beta_{f^\mu} = \frac{(1-X)g^2}{3(4\pi)^2}f^\mu. \tag{12}$$

However, in the theory with both c and f appearing in the Lagrangian, the coefficients cannot actually be measured independently. Up to the order we have considered, physical quantities such as the fermion energy-momentum relation will only depend on the particular linear combination $c_{\text{eff}}^{\nu\mu} = c^{\nu\mu} - f^\nu f^\mu/2$. According to (11–12), the RG flow for this physically observable combination is given by

$$\frac{\partial}{\partial(\log p/M)}\left(c^{\nu\mu} - \frac{1}{2}f^\nu f^\mu\right) = \frac{2g^2}{3(4\pi)^2}\left(c^{\nu\mu} - \frac{1}{2}f^\nu f^\mu\right), \tag{13}$$

independent of X. The physical quantity c_{eff} has the same RG evolution as c itself would have in a version of the theory with no f present at all.

So, the RG functions for certain operators in the SME are not unique and instead depend on the renormalization scheme, although not in a way that seems to lead to any direct inconsistencies. In fact, it is not unknown for RG β-functions to depend on the renormalization scheme even in conventional Lorentz-invariant QFTs. What is striking here, however, is that the scheme dependence arises already at one-loop order. Even at the classical level, the description of the fermion sector contains a redundancy, since it is possible to exchange an f coefficient for the equivalent c in tree-level calculations. It is not so surprising, then, that the redundancy has a manifestation at the quantum level as well.

3. Discrete Operators in the SME

Moreover, there is another manifestation of the ambiguity and the peculiar radiative corrections associated with it. The Lagrange density $\mathcal{L}_c$ is manifestly even under the discrete C and CPT symmetries. This follows from the usual transformation properties of the vector bilinear in the Dirac theory; normally, objects with even numbers of Lorentz indices are expected to be CPT-even. However, the equivalent theory described by $\mathcal{L}_f$ appears superficially to be CPT-odd, because it involves an operator with a single free index. The necessary resolution is that the physical discrete C, P, and T operators need to be transformed in accordance with (6). The operators no longer take their usual forms, but in the theory with f, they depend explicitly on the f coefficients, and this creates another interesting situation to explore.

For detailed calculations, we shall focus solely on the parity operator. The reasons for this are purely pragmatic. The actions of charge conjugation and time reversal involve complex conjugations, and so the matrix forms of the corresponding operators depend on the representation of the Dirac matrices. In contrast, the parity operator in the conventional theory is always simply γ^0, independently of how the Dirac matrices are chosen.

Another comment about the representation of the Dirac matrices is in order at this point. The field redefinitions that transfer Lorentz violation from the f to c operators may be reinterpreted as transformations of the Dirac matrices. In essence, a theory in which the only source of Lorentz violation is an f term is really a theory with a c term and a different choice of representation for the Clifford algebra of Dirac matrices.

The explicit matrix form of the parity operator in the $\mathcal{L}_f$ theory is[15]

$$\begin{aligned} \mathbf{P}\psi\mathbf{P} &= S_P\psi(t,-\vec{x}) \\ &= \left[\gamma^0 + i\frac{f^0}{\sqrt{1-f^2}}\gamma_5 + \frac{\left(1-\sqrt{1-f^2}\right)f^0 f^\nu}{f^2\sqrt{1-f^2}}\gamma_\nu\right]\psi(t,-\vec{x}). \end{aligned} \tag{14}$$

If the f is transformed away according to (6), S_P transforms into the usual form $S'_P = \gamma^0$. Although the form (14) is exact, it may be convenient to take the leading-order (in f) approximation, in light of the observed smallness of physical Lorentz violation,

$$S_P \approx \gamma^0 + if^0\gamma_5. \tag{15}$$

We can now make more concrete the previous observation that the operators parameterized by the f coefficients do not typically have well-defined behaviors under discrete spacetime symmetries. With the modified P operation (14), a spacelike f^j term in the Lagrange density is neither even nor odd under parity.[a]

What is potentially most consequential about the modified P (and C and T) is that it can have a nontrivial RG evolution. If the renormalization scheme is chosen with $X \neq 1$, so that $\beta_f \neq 0$, S_P will have the RG scale

[a]Thanks to S. Karki for this observation.

dependence

$$\frac{\partial}{\partial(\log p/M)}S_P = i\gamma_5\frac{\partial f^0}{\partial(\log p/M)} = i\gamma_5\frac{(1-X)g^2f^0}{3(4\pi)^2}. \tag{16}$$

Whether this is physically significant is unclear, but it is certainly a peculiar observation meriting more detailed analysis. There are many features of the SME at the quantum level that are not what we might have expected, based solely on experience with conventional Lorentz-invariant theories, and this is another such feature.

There is, however, at least one somewhat analogous situation that had definite importance—related to the tricky question of the proton spin and how relatively little of the total nucleon spin (or axial charge) may appear to derive from the spins of the underlying quarks. A fermion spin operator may naively appear to be purely a matrix operator Σ_0 in spinor space, therefore having no scale dependence. However, because of the chiral anomaly, the matrix elements of the total quark axial charge are modified to

$$\Sigma = \Sigma_0 - \frac{N_f\alpha_s}{2\pi}\Delta G, \tag{17}$$

and both the strong coupling α_s and the gluon polarization ΔG are scale-dependent quantities. While the anomalous scale dependence of the axial charge turned out to be a relatively minor part of the solution to the "proton spin crisis," it does provide a striking example where the scale dependence of a seemingly fixed matrix operator has direct physical consequences. It remains to be seen whether something similar may be said about the RG flow of the C, P, and T operators in the SME with a fermion f term.

Acknowledgments

I am grateful to S. Karki for his collaboration and his contributions to the understanding of the f-c duality.

References

1. D. Colladay and V.A. Kostelecký, *Phys. Rev. D* **55**, 6760 (1997).
2. D. Colladay and V.A. Kostelecký, *Phys. Rev. D* **58**, 116002 (1998).
3. O.W. Greenberg, *Phys. Rev. Lett.* **89**, 231602 (2002).
4. V.A. Kostelecký, C.D. Lane, and A.G.M. Pickering, *Phys. Rev. D* **65**, 056006 (2002).
5. G. de Berredo-Peixoto and I.L. Shapiro, *Phys. Lett. B* **642**, 153 (2006).
6. D. Colladay and P. McDonald, *Phys. Rev. D* **75**, 105002 (2007).
7. D. Colladay and P. McDonald, *Phys. Rev. D* **77**, 085006 (2008).

8. D. Colladay and P. McDonald, *Phys. Rev. D* **79**, 125019 (2009).
9. M. Gomes, J.R. Nascimento, A. Yu. Petrov, and A.J. da Silva, *Phys. Rev. D* **81**, 045018 (2010).
10. M.M. Anber and J.F. Donoghue, *Phys. Rev. D* **83**, 105027 (2011).
11. A. Ferrero and B. Altschul, *Phys. Rev. D* **84**, 065030 (2011).
12. L.C.T. Brito, J.C.C. Felipe, J.R. Nascimento, A. Yu. Petrov, and A.P. Baêta Scarpelli, *Phys. Rev. D* **102**, 075017 (2020).
13. B. Altschul, *J. Phys. A: Math. Gen.* **39**, 13757 (2006).
14. S. Karki and B. Altschul, *Eur. Phys. J. C* **82**, 676 (2022).
15. S. Karki and B. Altschul, *Symmetry* **14**, 2326 (2022).
16. S.D. Bass, *Rev. Mod. Phys.* **77**, 1257 (2005).
17. F. Myhrer and A.W. Thomas, *J. Phys. G: Nucl. Part. Phys.* **37**, 023101 (2010).

Chapter 17

Comments on the 2025 Data Tables for Lorentz and CPT Violation

N.E. Russell

Physics Department, Northern Michigan University, Marquette, MI 49855, USA

This contribution to the proceedings of the May 2025 Tenth Meeting on CPT and Lorentz Symmetry surveys recent limits placed on coefficients for Lorentz and CPT violation in the Standard-Model Extension framework since the ninth meeting held in 2022. Limits are organized based on the sector of the Standard Model or gravity in which they fall, the mass dimension of the relevant Lorentz-violating operators, and whether they are theoretically inferred or experimentally measured. The limits are all listed in the 2025 edition of the *Data Tables for Lorentz and CPT Violation*.

1. Introduction

The *Data Tables for Lorentz and CPT Violation*[1] record measured and derived values of coefficients for Lorentz and CPT violation. The 2025 update is the 18th edition, with 160 pages and 388 bibliographic references. The document contains three types of tabulations: summary tables, data tables, and properties tables. The bulk of the current document is the 48 data tables, which contain existing experimental measurements and limits derived indirectly by theoretical efforts using previously published experimental results. The constraints in the *Data Tables* are listed in the same form as in the literature with a minimum of editing, and in all cases the source citation is provided for the reader to delve more deeply. To provide some context, the experimental system is stated succinctly and citations to theoretically derived results are marked with an asterisk.

The Standard-Model Extension, or SME, has proven to be an essential framework for studies of Lorentz symmetry, and the *Data Tables* list the coefficients in this general system. It covers all the sectors of particle[2–4] and gravitational physics.[5,6] The SME Lagrange density function contains the operators for conventional physics, supplemented with Lorentz-breaking operators of all mass dimensions d constructed from conventional fields, coupled to coefficients for Lorentz violation.

In the quantum electrodynamics limit, the photon makes up the gauge sector, and matter is treated as three separate fermions: electrons, protons, and neutrons. The minimal SME involves operators of mass dimension $d = 3$ and $d = 4$, and the nonminimal SME involves higher dimensions and permits particle Lorentz-breaking couplings[3] between photons and fermions, as is seen in Table D23 of the current edition.

In the full Standard-Model limit, the SME Lagrange density includes operators for the charged-lepton sector, the neutrino sector, the quark sector, and the electroweak gauge sector.

The gravity sector of the SME introduces an additional level of complexity as couplings of the Standard-Model fields to gravitational fields such as the vierbein and the curvature tensor generate a plethora of operators. Among the technical challenges is the possibility of Lorentz-breaking theories that are not based in an underlying Riemann geometry.[7]

It should be noted that expressions for the Lagrange density in various useful limits can be found in the property tables section of the *Data Tables*. In addition, citations to the key theoretical papers are given in the explanatory text.

The goal of this contribution to the proceedings of the Tenth Meeting on CPT and Lorentz Symmetry is to summarize recent investigations of Lorentz symmetry by highlighting the data entries in the three *Data Tables* editions that have appeared since the previous meeting held in 2022. The list of references with recent additions includes 13 experimental citations and 23 theoretical citations, representing an increase in the number of references with SME bounds of about 10% over this period. In the following, these are discussed sector by sector.

2. Matter Sector

The 2025 *Data Tables* include 58 unique references with electron SME limits, 43 with proton limits, and 30 with neutron limits. In the last three editions, SME limits based on direct experiments have been found in hydrogen spectroscopy and trapped ions. On the theory front, results have been derived from earlier experiments with antihydrogen.

Hyperfine spectroscopy of a hydrogen beam has placed first limits on a variety of previously unbounded SME coefficients.[8] These are in the electron and proton sectors, and include coefficients in the minimal SME with mass dimension $d = 3$ and the nonminimal SME with mass dimensions $d = 5, 7$. The technique holds promise for similar results using antihydrogen. In another study, trapped Yb ions have yielded limits on the $d = 4$ minimal SME for the electron.[9]

A theory analysis of results from the ALPHA collaboration involving the $1S$–$2S$ transition in antihydrogen placed limits on electron and proton coefficients with mass dimensions $d = 3, 5$ and 7.[10]

3. Photon Sector

The 2025 edition of the *Data Tables* has 98 references with limits in the pure photon sector and 16 more in the photon-matter sector. Among these, seven were added in the last three editions.

A lower limit on the isotropic $d = 4$ photon-sector coefficient $\tilde{\kappa}_{\rm tr}$ has been found from LHC data considering photon decay.[11] In the astrophysics sphere, the spectral lag of gamma-ray bursts has been used to limit combinations of SME coefficients in the photon sector of dimension $d = 6$ and $d = 8$.[12] These limits cover a large selection of sources and are a useful complement to the existing limits.

Five theory-deduced limits have been added in the photon sector. Constraints on proton SME coefficients of mass dimensions $d = 3$ and $d = 4$ have been obtained through a theoretical analysis of Penning trap results looking at Lorentz- and CPT-violating effects at linear boost order.[13] CMB polarization data from Planck 2018 that measured the cosmic birefringence angle[14] have been used to place a theory-deduced limit on an isotropic $d = 3$ coefficient in the minimal photon sector.[15] This work is noteworthy due to the hint of a signal at a level above 2σ. A theoretical analysis of anisotropy in the cosmic microwave background has provided constraints on several photon-sector $d = 3, 4$ SME coefficients.[16] Satellite data measuring the solar wind over a period of six years was used to generate bounds on $d = 3$ photon coefficients in the SME.[17] An improved limit on a $d = 4$ photon-sector SME coefficient has been found from a theoretical analysis of ultra-high energy air showers.[18]

4. Charged Leptons

SME limits on electrons and positrons originate in 58 source documents in the 2025 edition of the *Data Tables*. Limits on the second- and third-generation charged leptons stem from 21 cited references, and these include 6 added in the last three editions.

A theoretical study found limits on isotropic $d = 4$ electron and muon SME coefficients based on the nonobservation of Cherenkov radiation of charged leptons and of photon decay into lepton-antilepton pairs.[19] Another theoretical study involving electric dipole moments placed limits on $d = 4$ SME coefficients for all three lepton generations.[20] In the muon sector, a study of electron-positron annihilation in the presence of strong magnetic fields has placed limits on $d = 4$ SME coefficients.[21] Limits on $d = 4$ muon- and tau-sector SME coefficients have been presented in a study of electron-positron scattering in strong magnetic fields.[22] Based on the SINDRUM II experiment at the Paul Scherrer Institute, a theoretical study of flavor-changing signatures of Lorentz and CPT violation has led to limits on $d = 5, 6$ SME coefficients in the lepton sector.[23] More than 500 constraints were placed on $d = 5$ SME coefficients in the lepton flavor-changing sector based on a theoretical analysis of experiments at the Paul Scherrer Institute and the Stanford Linear Accelerator.[24]

5. Neutrino Sector

Tables listing limits in the neutrino sector of the SME originate from 49 cited works, of which four were added in the last three editions of the *Data Tables*.

In the neutrino sector, the KATRIN collaboration has published limits on components of the $d = 3$ oscillation-free coefficients,[25] based on an analysis of tritium decay. The CUPID collaboration has limited oscillation-free $d = 3$ Lorentz violation by consideration of double beta decay in Mo atoms.[26] The IceCube collaboration published limits on SME coefficients in the neutrino sector for dimensions $d = 3$ through $d = 8$ based on more than seven years of data at their neutrino observatory in the Antarctic.[27] A theoretical study of flavor isotropy in IceCube data for e, μ, and τ neutrinos provided a limit on $d = 5$ coefficients in the neutrino sector.[28]

Publications in the neutrino sector have often provided large numbers of limits. In the 2025 edition of the *Data Tables*, 28 of the 108 pages with data tabulations are dedicated to the neutrino sector.

6. Quark Sector

In the current *Data Tables* edition, SME data in the quark sector are rooted in 32 cited documents. These include two experimental limits and three theory-deduced limits added in the last three editions.

The ZEUS Collaboration used deep inelastic scattering data from the HERA accelerator at DESY to find limits on quark coefficients in the SME. Their results include first limits on 30 coefficients and place bounds on $d = 4$ coefficients for the u, d, and s quarks and $d = 5$ coefficients for the u and s quarks.[29] The CMS collaboration at the LHC published limits on $d = 4$ quark coefficients based on $t\bar{t}$ production.[30]

In the category of theory-deduced constraints, a first limit on CPT violation in the top-quark sector based on LHC data has been placed on a $d = 3$ SME coefficient.[31] A theoretical study of Drell–Yan events at the LHC has limited $d = 4$ and $d = 5$ SME coefficients in the quark sector.[32] CPT-even coefficients of dimension $d = 4$ were bounded in another study of Drell–Yan events.[33]

7. Gauge Sector of the Minimal SME

The gauge sector of the 2025 *Data Tables* includes two tables, D43 and D44, and lists SME bounds from 11 sources in the electroweak category and two in the gluon category. This sector is relatively sparsely tested compared to other sectors of the SME. The large unexplored coefficient space means opportunities exist for useful results.

A theoretical analysis involving loop calculations of the anomalous magnetic dipole moment placed limits on SME coefficients of dimension $d = 3$ in the electroweak sector.[34] In another result, the impact of the electroweak gauge sector on the anomalous magnetic dipole moments of charged leptons was studied, and the results provide tight limits on $d = 3$ SME coefficients.[35]

8. Gravity Sector

The 2025 *Data Tables* document has 55 citations in the pure-gravity data tables and 47 in the gravity-fermion tables. In the last three years, three experimental citations and five theory-deduced citations were added in these sectors.

A Bragg interferometer involving Rb atoms forming a sensitive gravimeter was used to place limits on the $d = 4$ pure-gravity sector.[36] Limits on the $d = 5$ gravity sector of the SME were placed from an analysis of gravitational-wave dispersion from the GWTC-3 gravitational-wave transient catalog.[37] Free fall of Rb atoms in a dual-species atom interferometer has been used to limit SME coefficients in the $d = 3$ and $d = 5$ proton-gravity sector.[38] These proton-gravity coefficients control Lorentz violation operators in the SME Lagrange density that involve fermion fields coupled with gravitational fields.

In the category of theory-deduced limits, the nine $d = 4$ pure-gravity SME coefficients were bounded in a study of gravitational-wave events detected at two or more sites.[39] Strong bounds on the $d = 4$ gravity sector SME coefficients were placed by a theoretical study of pulsar timing data from a number of sources.[40] A theoretical study involving Sagittarius A* data from the Event Horizon Telescope has placed limits on $d = 4$ gravity sector coefficients.[41] A study of gravitational-wave propagation has placed limits on $d = 5$ and $d = 6$ SME coefficients in the nonminimal gravity sector.[42] A tight constraint on an isotropic $d = 5$ SME coefficient in the gravity sector has been placed in a theoretical study of results from the gravitational-wave catalogs GWTC-2 and GWTC-3.[43]

9. Closing

The number of unique citations in each sector of the *Data Tables* is shown in Fig. 1. The breakdown between experimental results and ones derived indirectly through theoretical considerations is provided for each case. More than half of the overall citations are in the QED sector, with the remainder split between the Standard-Model and gravity sectors. The highest experimental proportion is in the neutrino and photon-matter sectors, and the highest theory-deduced proportion is in the lepton, electroweak, and gluon sectors. Overall, the split between experimental and theory-derived bounds is approximately even.

Finding Lorentz violation in nature would be an unprecedented discovery in physics. The *Data Tables*, which provide a comprehensive collection of the experimental and theory-derived constraints on this profound symmetry, show no compelling evidence for Lorentz breaking. In a handful of cases signals have been found that rise above the level of 2σ, but this is to be expected statistically in a collection of thousands of measurements.

Several observations make the case for seeking evidence of Lorentz violations despite the growing body of evidence against it. Firstly, there is an infinite number of coefficients in the SME so there are many opportunities

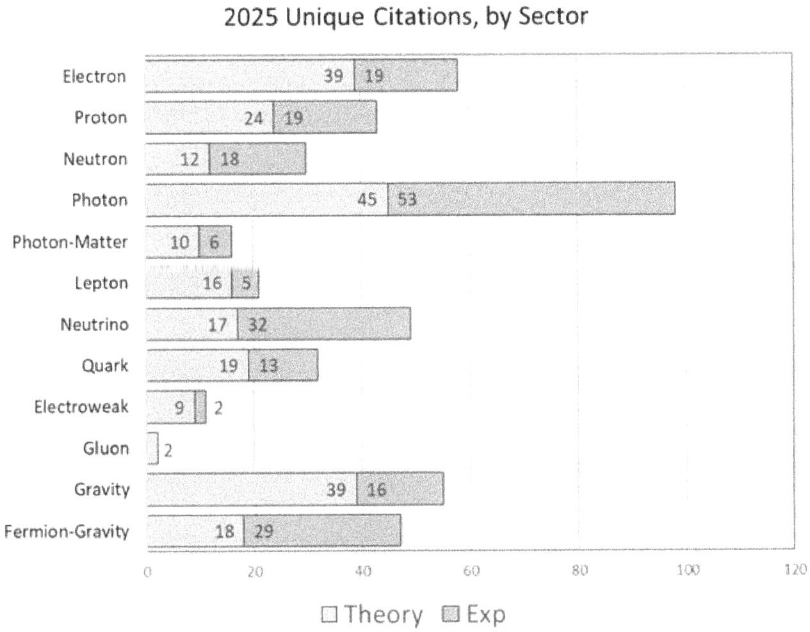

Fig. 1. Chart showing the number of distinct cited references in each sector of the 2025 *Data Tables for Lorentz and CPT Violation*. Each bar is subdivided to show the experimental and theory-derived citations.

to find this violation. Secondly, there are some coefficients where first measurements are yet to be made, and there are several sectors where only a few tests have been performed. For example, in the 2025 edition of the *Data Tables* all the results in the gluon sector are attributed to just two publications. Thirdly, the advancing frontier of experimental ingenuity and precision means that signals can appear in more sophisticated versions of earlier experiments. A fourth motivating factor is the scientific significance that broken Lorentz symmetry would carry. It would necessitate new theoretical developments and hold the promise of resolving issues such as the unification of the Standard Model and gravity.

References

1. V.A. Kostelecký and N. Russell, *Data Tables for Lorentz and CPT Violation*, 2025 edition, arXiv:0801.0287v18.
2. V.A. Kostelecký and R. Potting, *Phys. Rev. D* **51**, 3923 (1995).
3. D. Colladay and V.A. Kostelecký, *Phys. Rev. D* **55**, 6760 (1997).
4. D. Colladay and V.A. Kostelecký, *Phys. Rev. D* **58**, 116002 (1998).

5. V.A. Kostelecký, *Phys. Rev. D* **69**, 105009 (2004).
6. V.A. Kostelecký and Z. Li, *Phys. Rev. D* **103**, 024059 (2021).
7. V.A. Kostelecký and Z. Li, *Phys. Rev. D* **104**, 044054 (2021).
8. L. Nowak *et al.*, *Phys. Lett. B* **858**, 139012 (2024).
9. L.S. Dreissen, C.H. Yeh, H.A. Fürst, K.C. Grensemann, and T.E. Mehlstäubler, *Nat. Commun.* **13**, 7314 (2022).
10. A.J. Vargas, *Phys. Rev. D* **111**, 015032 (2025).
11. D. Amram, K. Bouzoud, N. Chanon, H. Hansen, M.R. Ribeiro, Jr., and M. Schreck, *Phys. Rev. Lett.* **132**, 211801 (2024).
12. J.N. Wei, Z.K. Liu, J.J. Wei, B.B. Zhang, and X.F. Wu, *Universe* **8**, 519 (2022).
13. A.J. Acevedo-López, Y. Ding, K. Iwasaki, and A.J. Vargas, *Phys. Rev. D* **108**, 076014 (2023).
14. Y. Minami and E. Komatsu, *Phys. Rev. Lett.* **125**, 221301 (2020).
15. N.A. Nilsson and C. Le Poncin-Lafitte, *Phys. Rev. D* **109**, 015032 (2024).
16. L. Caloni, S. Giardiello, M. Lembo, M. Gerbino, G. Gubitosi, M. Lattanzi, and L. Pagano, *JCAP* **03**, 018 (2023).
17. A.D.A.M. Spallicci, G. Sarracino, O. Randriamboarison, J.A. Helayël-Neto, and A. Dib, *Eur. Phys. J. Plus* **139**, 551 (2024).
18. F. Duenkel, M. Niechciol, and M. Risse, *Phys. Rev. D* **107**, 083004 (2023); PoS **ICRC2023**, 217 (2023); M. Niechciol, F. Duenkel, and M. Risse, *EPJ Web Conf.* **283**, 05007 (2023).
19. A. Crivellin, F. Kirk, and M. Schreck, *JHEP* **11**, 109 (2022).
20. S. Aghababaei, *Eur. Phys. J. C* **84**, 173 (2024).
21. M.H. Sis, *Phys. Lett. B* **848**, 138388 (2024).
22. D.S. Cabral and A.F. Santos, *Eur. Phys. J. Plus* **139**, 789 (2024).
23. V.A. Kostelecký, W.P. McNulty, E. Passemar, and N. Sherrill, *Phys. Lett. B* **868**, 139755 (2025).
24. V.A. Kostelecký, E. Passemar, and N. Sherrill, *Phys. Rev. D* **106**, 076016 (2022).
25. M. Aker *et al.*, KATRIN Collaboration, *Phys. Rev. D* **107**, 082005 (2023); C. Fengler, *PoS* **DISCRETE2022**, 011 (2024).
26. C. Augier *et al.*, CUPID Collaboration, *Eur. Phys. J. C* **84**, 925 (2024).
27. R. Abbasi *et al.*, IceCube Collaboration, *Nat. Phys.* **18**, 1287 (2022); C. Argüelles *et al.*, IceCube Collaboration, *PoS* **ICRC2023**, 1225 (2023).
28. B. Telalovic and M. Bustamante, *JCAP* **05**, 013 (2025).
29. I. Abt *et al.*, ZEUS Collaboration, *Phys. Rev. D* **107**, 092008 (2023).
30. A. Hayrapetyan *et al.*, CMS Collaboration, *Phys. Lett. B* **857**, 138979 (2024).
31. A. Belyaev, L. Cerrito, E. Lunghi, S. Moretti and N. Sherrill, *Phys. Rev. Lett.* **133**, 221601 (2024).
32. E. Lunghi and N. Sherrill, *Phys. Lett. B* **862**, 139366 (2025).
33. E. Lunghi, N. Sherrill, A. Szczepaniak, and A. Vieira, *JHEP* **04**, 228 (2021).
34. P.L. Cástulo, J.J. Toscano, and E.S. Tututi, *Int. J. Mod. Phys. A* **37**, 2250138 (2022).
35. J.S. Hurtado-Silva, J.J. Toscano, and O. Vázquez-Hernández, *J. Phys. G* **52**, 025002 (2025).

36. T. Zhang, L.L. Chen, Y.B. Shu, W.J. Xu, Y. Cheng, Q. Luo, Z.K. Hu, and M.K. Zhou, *Phys. Rev. Appl.* **20**, 014067 (2023).
37. L. Haegel, K. O'Neal-Ault, Q.G. Bailey, J.D. Tasson, M. Bloom, and L. Shao, *Phys. Rev. D* **107**, 064031 (2023); L. Haegel, *PoS* **ICHEP2022**, 076 (2022).
38. D. Gao, L. Zhou, J. Wang, and M. Zhan, *Phys. Rev. A* **110**, 043322 (2024).
39. A. Ray, P. Fan, V.F. He, M. Bloom, S.M. Yang, J.D. Tasson, and J.D.E. Creighton, *Phys. Rev. D* **110**, 122001 (2024).
40. Y. Dong, Z. Wang, and L. Shao, *Phys. Rev. D* **109**, 084024 (2024).
41. M. Khodadi and G. Lambiase, *Phys. Rev. D* **106**, 104050 (2022).
42. R. Niu, T. Zhu, and W. Zhao, *JCAP* **12**, 011 (2022).
43. Z.C. Zhao, Z. Cao, and S. Wang, *Astrophys. J.* **930**, 139 (2022).

Chapter 18

Flavor-Changing Signatures of Spacetime-Symmetry Violations

N. Sherrill

Institut für Theoretische Physik, Leibniz Universität Hannover, Appelstraße 2, 30167 Hannover, Germany

This proceedings contribution discusses Lorentz and CPT violation in the context of charged-lepton flavor-changing processes. Particular emphasis is placed on coherent muon-to-electron conversion in muonic atoms. Data collected by the SINDRUM II experiment at the Paul Scherrer Institute are used to extract constraints on mass dimension $d = 5$ electromagnetic and $d = 6$ quark-lepton effects. The Mu2e experiment at Fermilab and the COMET experiment at the Japan Proton Accelerator Complex are projected to improve constraints by up to two orders in magnitude.

1. Introduction

Charged-lepton flavor-changing (CLFC) processes initiated by Standard Model (SM) interactions in conjunction with neutrino oscillations yield practically unobservable effects.[1] Experiments in search of CLFC processes thus provide clean probes of physics beyond the SM.[2–4] This contribution examines CLFC processes initiated by small violations of Lorentz and CPT invariance, with an emphasis on neutrinoless and coherent muon-to-electron conversion in muonic atoms.[5]

A model-independent description of Lorentz violation (LV) and other fundamental symmetries is provided by the Standard-Model Extension (SME) effective field theory.[6–9] The SME contains operators describing CLFC interactions, of which only a subset involving muon decays has been investigated.[6,10–20] Using existing data collected by the SINDRUM II experiment at the Paul Scherrer Institute,[21] we revisit constraints on mass dimension $d = 5$ operators first extracted from electromagnetic decays[20] and deduce several first constraints on $d = 6$ quark-lepton operators.

One golden CLFC channel is the coherent conversion $\mu + N \to e + N$ of a muon μ into an electron e in the presence of a nucleus N. Experiments searching for this process involve a proton beam that strikes a production target, generating pions and other hadrons that decay to muons among other products. The muons are further directed onto a stopping target, forming muonic atoms that relax to the 1S ground state. The signal for a neutrinoless muon-electron conversion is a monoenergetic electron ejected from a muonic atom with energy $E_e^{\text{conv}} = m_\mu - E_{\text{bind}} - E_{\text{recoil}}$, where m_μ is the muon mass, E_{bind} is the muon binding energy, and E_{recoil} is the nuclear recoil energy. This process can be characterized experimentally by the dimensionless ratio

$$R_{\mu e} = \frac{\omega_{\text{conv}}}{\omega_{\text{capt}}}, \tag{1}$$

where ω_{conv} and ω_{capt} are the muon conversion and capture rates, respectively. The dominant contribution to ω_{conv} is the coherent conversion $\mu^- + {}^A_Z N \to e^- + {}^A_Z N$, where the nucleus ${}^A_Z N$ of the muonic atom has atomic number A and charge Z.[22]

To date, SINDRUM II has achieved the most stringent constraint, $R_{\mu e} < 7 \times 10^{-13}$ at the 90% confidence level (CL), using ${}^{197}_{79}\text{Au}$ as the target ($\omega_{\text{capt}}^{\text{Au}} \simeq 13.07$ MHz[23]).[21] Upcoming searches using an ${}^{27}_{13}\text{Al}$ target ($\omega_{\text{capt}}^{\text{Al}} \simeq 0.7054$ MHz[23]) include the Muon-to-Electron Conversion (Mu2e) experiment at Fermilab, which anticipates achieving $R_{\mu e} \simeq 6.2 \times 10^{-16}$,[24] and the Coherent Muon to Electron Transition (COMET) experiment at the Japan Proton Accelerator Complex (J-PARC), which expects to reach $R_{\mu e} \simeq 7 \times 10^{-15}$.[25] More distant future upgrades include phase two of Mu2e,[26] and phase two of COMET[27] and other experiments at J-PARC, which expect to reach $R_{\mu e} \simeq 10^{-18}$.[28]

No observable CLFC effects in muon decay involve SME terms of mass dimension $d = 3$ or 4.[20] The leading contributions instead arise via $d = 5$ electromagnetic interactions and $d = 6$ (4-point) quark-lepton interactions rather than from propagator terms.[29] Since the 4-point interactions involve the quarks in the nucleus, the corresponding CLFC effects are uniquely accessed in the coherent conversion $\mu + N \to e + N$. We report the first constraints on these latter effects.[5]

In the following sections, we calculate the leading SME contributions to the conversion rate $\omega_{\rm conv}$ for muonic gold and aluminum targets. Constraints are extracted from the SINDRUM II result,[21] and projected constraints are estimated for the upcoming Mu2e and COMET experiments.

2. Setup

The muon and electron wavefunctions are written as

$$\psi_{\kappa,s}(r,\theta,\phi) = \begin{pmatrix} g(r)\chi_{\kappa,s}(\theta,\phi) \\ if(r)\chi_{-\kappa,s}(\theta,\phi) \end{pmatrix}, \tag{2}$$

where $\kappa = \mp(J+1/2)$ and the eigenspinors $\chi_{\kappa,s}$ satisfy $(\boldsymbol{\sigma}\cdot\boldsymbol{l}+I)\chi_{\kappa,s} = -\kappa\chi_{\kappa,s}$ and $J_z\chi_{\kappa,s} = s\chi_{\kappa,s}$, with normalization $\int d\Omega\, \chi^\dagger_{\kappa',s'}\chi_{\kappa,s} = \delta_{s's}\delta_{\kappa'\kappa}$. The 1S muon wavefunction has $\kappa = -1$, and neglecting the electron mass implies $g^+ = if^-$ and $f^+ = -ig^-$. The radial wavefunctions are obtained by numerically solving the Dirac equation

$$\frac{d}{dr}\begin{pmatrix} u_1 \\ u_2 \end{pmatrix} = \begin{pmatrix} -\kappa/r & W - V(r) + m \\ -(W - V(r) - m) & \kappa/r \end{pmatrix}\begin{pmatrix} u_1 \\ u_2 \end{pmatrix} \tag{3}$$

for $u_1(r) = rg(r)$ and $u_2(r) = rf(r)$. Here, W is the lepton energy and $V(r)$ is the approximately static and spherically symmetric nuclear electric potential.[30] The potential is obtained from $\rho(r)$, which is normalized as $\int_0^\infty 4\pi\rho(r)r^2\,dr = Z$. The form of $\rho(r)$ is taken as a two-parameter Fermi model for $^{197}_{79}$Au and as a Fourier-Bessel expansion for $^{27}_{13}$Al.[31,32]

The conversion rate for $d=5$ electromagnetic interactions may be written as

$$\omega_{\rm conv} = \tfrac{1}{2}\sum_{s=\pm\frac{1}{2}}\sum_{\kappa=\pm 1}\sum_{s'=\pm\frac{1}{2}}\left|\int d^3x F_{\alpha\beta}\bar{\psi}^{(e)}_{\kappa,s'}\mathcal{O}^{\alpha\beta}\psi^{(\mu)}_s\right|^2, \tag{4}$$

where $F_{\alpha\beta}$ is the electromagnetic field strength, $\psi^{(e)}_{\kappa,s'}$ is a continuum electron wavefunction, and $\psi^{(\mu)}_s$ is the 1S bound-state muon wavefunction. The matrix $\mathcal{O}^{\alpha\beta}$ contains the set $\{(m^{(5)}_F)^{\alpha\beta}_{\mu e},\ i(m^{(5)}_{5F})^{\alpha\beta}_{\mu e}\gamma_5,\ (a^{(5)}_F)^{\lambda\alpha\beta}_{\mu e}\gamma_\lambda,\ (b^{(5)}_F)^{\lambda\alpha\beta}_{\mu e}\gamma_5\gamma_\lambda,\ \tfrac{1}{2}(H^{(5)}_F)^{\kappa\lambda\alpha\beta}_{\mu e}\sigma_{\kappa\lambda}\}$.[8,20] These modify interactions rather than propagators, so standard perturbative quantization methods hold[33] and Eq. (4) can be derived using tree-level Feynman rules.

The conversion rate for the $d = 6$ quark-lepton interactions may be written as

$$\omega_{\text{conv}} = \tfrac{1}{2}\sum_{s,s',\kappa} \left| \int d^3x \left(\alpha \bar{\psi}^{(e)}_{\kappa,s'} \mathcal{K} \psi^{(\mu)}_s + \beta \bar{\psi}^{(e)}_{\kappa,s'} \mathcal{K}_0 \psi^{(\mu)}_s \right) \right|^2, \tag{5}$$

where α and β are conventional nuclear matrix elements given by $\alpha = \langle N|\bar{\psi}^{(q)}\psi^{(q)}|N\rangle$ with q summed over the quark flavors $q = u, d, s$ and $\beta = \langle N|\bar{\psi}^{(q)}\gamma_0\psi^{(q)}|N\rangle$ with q summed over $q = u, d$.[34] Other possible nuclear matrix elements vanish in coherent conversion.[32] The matrices $\mathcal{K}$ and $\mathcal{K}_0$ incorporate the $d = 6$ SME coefficients,[8] where $\mathcal{K}$ contains the set $\{(k^{(6)}_{SV})^{\lambda}_{qqe\mu}\gamma_\lambda,\ (k^{(6)}_{SA})^{\lambda}_{qqe\mu}\gamma_5\gamma_\lambda,\ \tfrac{1}{2}(k^{(6)}_{ST})^{\kappa\lambda}_{qqe\mu}\sigma_{\kappa\lambda}\}$ for $q = u, d, s$ and $\mathcal{K}_0$ contains the set $\{(k^{(6)}_{VS})^{t}_{qqe\mu},\ i(k^{(6)}_{VP})^{t}_{qqe\mu}\gamma_5,\ (k^{(6)}_{VV})^{t\lambda}_{qqe\mu}\gamma_\lambda,\ (k^{(6)}_{VA})^{t\lambda}_{qqe\mu}\gamma_5\gamma_\lambda,\ \tfrac{1}{2}(k^{(6)}_{VT})^{t\kappa\lambda}_{qqe\mu}\sigma_{\kappa\lambda}\}$ for $q = u, d$.

3. Experiments and Constraints

For the experiments considered here, the cartesian SME coefficients appearing in $\mathcal{O}^{\alpha\beta}$, $\mathcal{K}$, and $\mathcal{K}_0$ can be taken to be independent of time and location.[7] The coefficients change under observer Lorentz transformations, implying measurements of their values must be provided in a specified inertial frame. The canonical choice is the Sun-centered frame (SCF) with cartesian coordinates (T, X, Y, Z). The time origin $T = 0$ is defined as the 2000 vernal equinox, the X axis is chosen to point from the Earth to the Sun at $T = 0$, and the Z axis is parallel to the Earth's rotation axis.[35] To establish the transformation from the SCF to the laboratory detector frame (DF), it is also useful to introduce a standard Earth-based laboratory frame (LF) with cartesian coordinates (x, y, z) and the $x(y)$ axis pointing to local south (east).

Laboratory frames are typically noninertial due to the rotation of the Earth at the sidereal frequency $\omega_\oplus \simeq 2\pi/(23\,\text{h}\ 56\,\text{min})$. At leading order, the transformation from the SCF to the LF is a rotation $\mathcal{R}(\chi, \omega_\oplus T_\oplus)$ that depends on the laboratory colatitude χ and harmonics[36] of the laboratory sidereal time $T_\oplus \equiv T - T_0$. The time $T_\oplus$ in the LF is shifted by an experiment-dependent amount T_0 relative to the time T in the SCF, arising from the laboratory longitude λ and other effects.[37] The explicit form of the rotation $\mathcal{R}(\chi, \omega_\oplus T_\oplus)$ is given as Eq. (7) of Ref. 20. The DF typically differs from the standard LF, so we must perform an additional transformation $\mathcal{R}_{\text{detector}}(\psi)$ given in Eq. (6) of Ref. 20 that involves the angle ψ of the laboratory z-axis measured north of east along the direction of the beamline. The net transformation from the SCF to the DF is thus $\mathcal{R}_{\text{total}} = \mathcal{R}_{\text{detector}}(\psi)\mathcal{R}(\chi, \omega_\oplus T_\oplus)$. Applying the transformation $\mathcal{R}_{\text{total}}$ yields $\omega_{\text{conv}} = \omega_{\text{conv}}(\psi, \chi, \omega_\oplus T_\oplus)$, and thus $R_{\mu e} = R_{\mu e}(\psi, \chi, \omega_\oplus T_\oplus)$, in terms of the detector orientation, detector location, sidereal harmonics, and the SME coefficients in the SCF. The quantities χ, λ, T_0, and ψ for the experiments considered here are listed in Table 1.

The spherical symmetry of the nuclear charge distribution implies contributions to Eq. (4) arise only from the field strength components $F_{tr} = -F_{rt} = E_r(r)$, where the index t represents the time $T_\oplus$. As $\mathcal{O}^{tr}$ is a spatial vector, the match between polar and cartesian coordinates in the DF can be accomplished by $\mathcal{O}^{tr} = \mathcal{O}^{tx} \sin\theta\cos\phi + \mathcal{O}^{ty} \sin\theta\sin\phi + \mathcal{O}^{tz}\cos\theta$, as usual. Working in DF spherical-polar coordinates with the z axis directed along the muon beam direction, the azimuthal integral can then be evaluated over the full range $\phi \in [0, 2\pi)$. In contrast, the polar angle θ is limited by the detector acceptance to a range $|\cos\theta| \lesssim c$. Integration over θ results in two geometrical factors $w_{xy} = -c^3/6 + c/2$ and $w_z = c^3/3$. The approximate values of the acceptances c are listed in Table 2.

The radial part of Eq. (4) gives two integrals unique to each experiment,

$$\begin{aligned} I_1 &= \int dr\, r^2 E(r) \left(f_e^- g_\mu^- + g_e^- f_\mu^-\right), \\ I_2 &= \int dr\, r^2 E(r) \left(g_e^- f_\mu^- - f_e^- g_\mu^-\right). \end{aligned} \tag{6}$$

Table 2 displays numerical values of the quantities $\widetilde{I}_1 = I_1/m_\mu^{3/2}$ and $\widetilde{I}_2 = I_2/m_\mu^{3/2}$. Combining these results gives the conversion rate ω_{conv} expressed using SME coefficients defined in the DF.

Table 1. Detector-frame parameters.

Experiment	χ	λ	T_0	ψ
SINDRUM II	42.5°	8.2°	3.86 h	242°
COMET	53.6°	140.6°	−4.94 h	188°
Mu2e	48.2°	−88.2°	10.27 h	122°

Table 2. Detector acceptances and normalized radial integrals.

Experiment	c	$\widetilde{I}_1$	$\widetilde{I}_2$	$\widetilde{I}_3$	$\widetilde{I}_4$
SINDRUM II	0.44	−0.053	0.070	0.040	0.18
COMET	0.34	−0.012	0.012	7.6×10^{-4}	0.041
Mu2e	0.50	−0.012	0.012	7.6×10^{-4}	0.041

Time averaging can be performed for data collected over extended intervals, leaving only the time-independent contribution $\overline{\omega}_{\rm conv}$. Following accepted procedure, an experimental measurement of $\overline{\omega}_{\rm conv}$ can thereby be translated into constraints on the components of the SME coefficients taken one at a time. The SINDRUM II result $R_{\mu e} < 7\times 10^{-13}$ was obtained using a primary dataset taken over 81 days.[21] It can therefore be interpreted as a limit on the time-averaged conversion rate $\overline{\omega}_{\rm conv} = \overline{\omega}_{\rm conv}(\psi, \chi)$ for the corresponding ψ and χ values given in Table 1.

With $R_{\mu e}$, $\omega_{\rm capt}$, and $\overline{\omega}_{\rm conv}$, constraints on the $d = 5$ SME coefficient components are extracted and presented in Table 3.

Each result is a constraint at 90% CL on a coefficient component in the SCF, where the indices J and $K \neq J$ are X or Y. The SINDRUM II result implies sensitivities at parts in 10^{12} to 96 real components. The constraints involving the coefficients $(a_F^{(5)})_{\mu e}^{\lambda\alpha\beta}$ and $(b_F^{(5)})_{\mu e}^{\lambda\alpha\beta}$ also represent limits on CPT violation.

Table 3. Constraints on $d = 5$ SME coefficients in units of 10^{-12} GeV^{-1}. The results for SINDRUM II are constraints at 90% CL. The results for Mu2e and COMET are projected constraints based on expected rates.

Coefficients	SINDRUM II	COMET	Mu2e
$\lvert(m_F^{(5)})_{\mu e}^{TJ}\rvert, \lvert(m_{5F}^{(5)})_{\mu e}^{TJ}\rvert$	< 8	< 1	< 0.2
$\lvert(m_F^{(5)})_{\mu e}^{TZ}\rvert, \lvert(m_{5F}^{(5)})_{\mu e}^{TZ}\rvert$	< 8	< 0.9	< 0.2
$\lvert(a_F^{(5)})_{\mu e}^{TTJ}\rvert, \lvert(b_F^{(5)})_{\mu e}^{TTJ}\rvert$	< 6	< 1	< 0.2
$\lvert(a_F^{(5)})_{\mu e}^{TTZ}\rvert, \lvert(b_F^{(5)})_{\mu e}^{TTZ}\rvert$	< 6	< 0.9	< 0.2
$\lvert(a_F^{(5)})_{\mu e}^{JTJ}\rvert, \lvert(b_F^{(5)})_{\mu e}^{JTJ}\rvert$	< 6	< 1	< 0.2
$\lvert(a_F^{(5)})_{\mu e}^{JTK}\rvert, \lvert(b_F^{(5)})_{\mu e}^{JTK}\rvert$	< 8	< 1	< 0.2
$\lvert(a_F^{(5)})_{\mu e}^{JTZ}\rvert, \lvert(b_F^{(5)})_{\mu e}^{JTZ}\rvert$	< 8	< 0.9	< 0.2
$\lvert(a_F^{(5)})_{\mu e}^{ZTJ}\rvert, \lvert(b_F^{(5)})_{\mu e}^{ZTJ}\rvert$	< 8	< 1	< 0.2
$\lvert(a_F^{(5)})_{\mu e}^{ZTZ}\rvert, \lvert(b_F^{(5)})_{\mu e}^{ZTZ}\rvert$	< 7	< 0.9	< 0.2
$\lvert(H_F^{(5)})_{\mu e}^{TJTJ}\rvert, \lvert(H_F^{(5)})_{\mu e}^{JZTK}\rvert$	< 7	< 1	< 0.2
$\lvert(H_F^{(5)})_{\mu e}^{TJTK}\rvert, \lvert(H_F^{(5)})_{\mu e}^{JZTJ}\rvert$	< 6	< 1	< 0.2
$\lvert(H_F^{(5)})_{\mu e}^{TJTZ}\rvert, \lvert(H_F^{(5)})_{\mu e}^{JZTZ}\rvert$	< 6	< 0.9	< 0.2
$\lvert(H_F^{(5)})_{\mu e}^{TZTJ}\rvert, \lvert(H_F^{(5)})_{\mu e}^{XYTJ}\rvert$	< 6	< 1	< 0.2
$\lvert(H_F^{(5)})_{\mu e}^{TZTZ}\rvert, \lvert(H_F^{(5)})_{\mu e}^{XYTZ}\rvert$	< 7	< 0.9	< 0.2

Table 4. Constraints on $d = 6$ SME coefficients in units of 10^{-13} GeV^{-2}. The results for SINDRUM II are constraints at 90% CL. The results for Mu2e and COMET are projected constraints based on expected rates.

Coefficients	SINDRUM II	COMET	Mu2e
$\lvert(k_{SV}^{(6)})_{uue\mu}^{T}\rvert, \lvert(k_{SA}^{(6)})_{uue\mu}^{T}\rvert$	< 6	< 1	< 0.2
$\lvert(k_{SV}^{(6)})_{uue\mu}^{J}\rvert, \lvert(k_{SA}^{(6)})_{uue\mu}^{J}\rvert$	< 7	< 1	< 0.2
$\lvert(k_{SV}^{(6)})_{uue\mu}^{Z}\rvert, \lvert(k_{SA}^{(6)})_{uue\mu}^{Z}\rvert$	< 7	< 1	< 0.2
$\lvert(k_{SV}^{(6)})_{dde\mu}^{T}\rvert, \lvert(k_{SA}^{(6)})_{dde\mu}^{T}\rvert$	< 6	< 1	< 0.2
$\lvert(k_{SV}^{(6)})_{dde\mu}^{J}\rvert, \lvert(k_{SA}^{(6)})_{dde\mu}^{J}\rvert$	< 7	< 1	< 0.2
$\lvert(k_{SV}^{(6)})_{dde\mu}^{Z}\rvert, \lvert(k_{SA}^{(6)})_{dde\mu}^{Z}\rvert$	< 7	< 1	< 0.2
$\lvert(k_{SV}^{(6)})_{sse\mu}^{T}\rvert, \lvert(k_{SA}^{(6)})_{sse\mu}^{T}\rvert$	< 10	< 2	< 0.4
$\lvert(k_{SV}^{(6)})_{sse\mu}^{J}\rvert, \lvert(k_{SA}^{(6)})_{sse\mu}^{J}\rvert$	< 15	< 2	< 0.4
$\lvert(k_{SV}^{(6)})_{sse\mu}^{Z}\rvert, \lvert(k_{SA}^{(6)})_{sse\mu}^{Z}\rvert$	< 15	< 2	< 0.4
$\lvert(k_{VS}^{(6)})_{uue\mu}^{T}\rvert, \lvert(k_{VP}^{(6)})_{uue\mu}^{T}\rvert$	< 30	< 4	< 0.8
$\lvert(k_{VS}^{(6)})_{dde\mu}^{T}\rvert, \lvert(k_{VP}^{(6)})_{dde\mu}^{T}\rvert$	< 30	< 4	< 0.7
$\lvert(k_{ST}^{(6)})_{uue\mu}^{TJ}\rvert, \lvert(k_{ST}^{(6)})_{uue\mu}^{JZ}\rvert$	< 7	< 1	< 0.2
$\lvert(k_{ST}^{(6)})_{uue\mu}^{TZ}\rvert, \lvert(k_{ST}^{(6)})_{uue\mu}^{XY}\rvert$	< 7	< 1	< 0.2
$\lvert(k_{ST}^{(6)})_{dde\mu}^{TJ}\rvert, \lvert(k_{ST}^{(6)})_{dde\mu}^{JZ}\rvert$	< 7	< 1	< 0.2
$\lvert(k_{ST}^{(6)})_{dde\mu}^{TZ}\rvert, \lvert(k_{ST}^{(6)})_{dde\mu}^{XY}\rvert$	< 7	< 1	< 0.2
$\lvert(k_{ST}^{(6)})_{sse\mu}^{TJ}\rvert, \lvert(k_{ST}^{(6)})_{sse\mu}^{JZ}\rvert$	< 15	< 2	< 0.4
$\lvert(k_{ST}^{(6)})_{sse\mu}^{TZ}\rvert, \lvert(k_{ST}^{(6)})_{sse\mu}^{XY}\rvert$	< 15	< 2	< 0.4
$\lvert(k_{VV}^{(6)})_{uue\mu}^{TT}\rvert, \lvert(k_{VA}^{(6)})_{uue\mu}^{TT}\rvert$	< 20	< 4	< 0.7
$\lvert(k_{VV}^{(6)})_{uue\mu}^{TJ}\rvert, \lvert(k_{VA}^{(6)})_{uue\mu}^{TJ}\rvert$	< 25	< 4	< 0.7
$\lvert(k_{VV}^{(6)})_{uue\mu}^{TZ}\rvert, \lvert(k_{VA}^{(6)})_{uue\mu}^{TZ}\rvert$	< 25	< 4	< 0.7
$\lvert(k_{VV}^{(6)})_{dde\mu}^{TT}\rvert, \lvert(k_{VA}^{(6)})_{dde\mu}^{TT}\rvert$	< 20	< 4	< 0.7
$\lvert(k_{VV}^{(6)})_{dde\mu}^{TJ}\rvert, \lvert(k_{VA}^{(6)})_{dde\mu}^{TJ}\rvert$	< 20	< 4	< 0.7
$\lvert(k_{VV}^{(6)})_{dde\mu}^{TZ}\rvert, \lvert(k_{VA}^{(6)})_{dde\mu}^{TZ}\rvert$	< 20	< 4	< 0.7
$\lvert(k_{VT}^{(6)})_{uue\mu}^{TTJ}\rvert, \lvert(k_{VT}^{(6)})_{uue\mu}^{TJZ}\rvert$	< 25	< 4	< 0.7
$\lvert(k_{VT}^{(6)})_{uue\mu}^{TTZ}\rvert, \lvert(k_{VT}^{(6)})_{uue\mu}^{TXY}\rvert$	< 25	< 4	< 0.7
$\lvert(k_{VT}^{(6)})_{dde\mu}^{TTJ}\rvert, \lvert(k_{VT}^{(6)})_{dde\mu}^{TJZ}\rvert$	< 20	< 4	< 0.7
$\lvert(k_{VT}^{(6)})_{dde\mu}^{TTZ}\rvert, \lvert(k_{VT}^{(6)})_{dde\mu}^{TXY}\rvert$	< 20	< 4	$< 0.7.$

Following a similar approach for the $d = 6$ quark-lepton conversion rate (5) results in two radial integrals

$$I_3 = \int dr\, r^2 \rho^{(p)}(r) f_e^- f_\mu^-, \quad I_4 = \int dr\, r^2 \rho^{(p)}(r) g_e^- g_\mu^-. \tag{7}$$

The numerical values of $\widetilde{I}_3 = I_3/m_\mu^{5/2}$ and $\widetilde{I}_4 = I_4/m_\mu^{5/2}$ are given in Table 2.

Again interpreting the SINDRUM II result as a constraint on the time-averaged conversion rate $\overline{\omega}_{\rm conv}$, we establish and present in Table 4 first constraints on each of the $d = 6$ coefficient components in the SCF, where J is X or Y. The results show that SINDRUM II probes 148 real coefficient components arising from 4-point CLFC quark-lepton interactions to parts in 10^{13}. The constraints for components with one or three spacetime indices also represent constraints on CPT violation. Table 4 also displays the corresponding projections for the Mu2e and COMET experiments.

Together with the enhanced resolving power of a sidereal analysis, we suggest that conversion experiments at Fermilab and J-PARC could ultimately yield orders-of-magnitude improvements over the present results. This provides an optimistic outlook for future tests of Lorentz and CPT violation in CLFC processes.

Acknowledgments

This work was supported by the Deutsche Forschungsgemeinschaft under the Heinz Maier Leibnitz Prize BeyondSM HML-537662082, and by the Indiana University Center for Spacetime Symmetries.

References

1. W.J. Marciano and A.I. Sanda, *Phys. Lett. B* **67**, 303 (1977); S.M. Bilenky, S.T. Petcov, and B. Pontecorvo, *Phys. Lett. B* **67**, 309 (1977); B.W. Lee, S. Pakvasa, R.E. Schrock, and H. Sugawara, *Phys. Rev. Lett.* **38**, 937 (1977).
2. R. Barbieri and L.J. Hall, *Phys. Lett. B* **338**, 212 (1994).
3. L. Calibbi and G. Signorelli, *Riv. Nuovo Cim.* **41**, 71 (2018).
4. R.H. Bernstein and P.S. Cooper, *Phys. Rept.* **532**, 27 (2013).
5. A. Kostelecký, W.P. McNulty, E. Passemar, and N. Sherrill, *Phys. Lett. B* **868**, 139755 (2025).
6. D. Colladay and V.A. Kostelecký, *Phys. Rev. D* **55**, 6760 (1997); *Phys. Rev. D* **58**, 116002 (1998).
7. V.A. Kostelecký, *Phys. Rev. D* **69**, 105009 (2004).
8. V.A. Kostelecký and Z. Li, *Phys. Rev. D* **99**, 056016 (2019); *Phys. Rev. D* **103**, 024059 (2021).

9. For reviews see, for example, Q.G. Bailey, in C. Bambi and A. Cardenas-Avendano, eds., *Recent Progress on Gravity Tests*, Springer, Singapore, 2024; Á. Roberts, Galaxies **9**, 47 (2021); J.D. Tasson, *Rep. Prog. Phys.* **77**, 062901 (2014); R. Bluhm, *Lect. Notes Phys.* **702**, 191 (2006).
10. V.A. Kostelecký and N. Russell, *Data Tables for Lorentz and CPT Violation*, *Rev. Mod. Phys.* **83**, 11 (2011), arXiv:0801.0287v18 (2025).
11. S.R. Coleman and S.L. Glashow, *Phys. Rev. D* **59**, 116008 (1998).
12. E.O. Iltan, JHEP **6**, 016 (2003).
13. Y.M.P. Gomes and J.A. Helayël-Neto, *Eur. Phys. J. C* **80**, 287 (2020).
14. A. Crivellin, F. Kirk, and M. Schreck, *JHEP* **4**, 82 (2021).
15. S. Hollenberg and P.B. Pal, *Phys. Lett. B* **701**, 89 (2011).
16. J.P. Noordmans, C.J.G. Onderwater, H.W. Wilschut, and R.G.E. Timmermans, *Phys. Rev. D* **93**, 116001 (2016).
17. M.A. López-Osorio, E. Martínez-Pascual, and J.J. Toscano, *J. Phys. G* **43**, 025003 (2016).
18. V.E. Mouchrek-Santos and M.M. Ferreira, Jr., *Phys. Rev. D* **95**, 071701 (2017).
19. Y.M.P. Gomes, P.C. Malta, and M.J. Neves, *Phys. Rev. D* **101**, 075001 (2020).
20. V.A. Kostelecký, E. Passemar, and N. Sherrill, *Phys. Rev. D* **106**, 076016 (2022).
21. W.H. Bertl *et al.*, *Eur. Phys. J. C* **47**, 337 (2006); A. van der Schaaf, *SciPost Phys. Proc.* **5**, 8 (2021).
22. A. Czarnecki, W.J. Marciano, and K. Meinikov, hep-ph/9801218; H.C. Chiang, E. Oset, T.S. Kosmas, A. Faessler, and J.D. Vergados, *Nucl. Phys. A* **559**, 526 (1993).
23. T. Suzuki, D.F. Measday, and J.P. Roalsvig, *Phys. Rev. C* **35**, 2212 (1987).
24. L. Bartoszek *et al.*, arXiv:1501.05241; K. Byrum *et al.*, *Universe* **2023**, 54 (2023).
25. R. Abramishvili *et al.*, *PTEP* **2020**, 033C01 (2020).
26. K. Byrum *et al.*, arXiv:2203.07569.
27. A. Kurup, *PoS* **NUFACT2014**, 101 (2014); B.E. Krikler, Ph.D. thesis, Imperial College, London, England (2017).
28. M. Aoki *et al.*, Letter of Intent P20-LoI, First PAC Meeting at J-PARC, June 2006; Y. Kuno, *Nucl. Phys. B Proc. Suppl.* **149**, 376 (2005).
29. A.H. Gomes, V.A. Kostelecký, and A.J. Vargas, *Phys. Rev. D* **90**, 076009 (2014); R. Bluhm, V.A. Kostelecký, and C.D. Lane, *Phys. Rev. Lett.* **84**, 1098 (2000).
30. P. Möller, A.J. Sierk, T. Ichikawa, and H. Sagawa, *Atom. Data Nucl. Data Tables* **109**, 1 (2016); M. Haouas, F. Taulelle, and C. Martineau, *Prog. Nucl. Magn. Reson. Spectr.* **94**, 11 (2016); B. Bally, G. Giacalone, and M. Bender, *Eur. Phys. J. A* **59**, 58 (2023).
31. H. de Vries, C.W. de Jager, and C. de Vries, At. *Data Nucl. Data Tables* **36**, 495 (1987).
32. R. Kitano, M. Koike, and Y. Okada, *Phys. Rev. D* **66**, 096002 (2002).

33. V.A. Kostelecký and R. Lehnert, *Phys. Rev. D* **63**, 065008 (2001); V.A. Kostelecký, R. Lehnert, M. Schreck, and B. Seradjeh, *New J. Phys.* **27**, 073901 (2025).
34. T.S. Kosmas, S. Kovalenko, and I. Schmidt, *Phys. Lett. B* **511**, 203 (2001).
35. R. Bluhm, V.A. Kostelecký, C.D. Lane, and N. Russell, *Phys. Rev. D* **68**, 125008 (2003); *Phys. Rev. Lett.* **88**, 090801 (2002); V.A. Kostelecký and M. Mewes, *Phys. Rev. D* **66**, 056005 (2002).
36. V.A. Kostelecký, *Phys. Rev. Lett.* **80**, 1818 (1998).
37. V.A. Kostelecký, A.C. Melissinos, and M. Mewes, *Phys. Lett. B* **761**, 1 (2016); Y. Ding and V.A. Kostelecký, *Phys. Rev. D* **94**, 056008 (2016).

Chapter 19

Vacuum Cherenkov Radiation for Nonminimal Isotropic Lorentz Violation

Albert Yu. Petrov,[*] Marco Schreck,[†,‡] and Alexandre R. Vieira[§]

[*]*Departamento de Física, Universidade Federal da Paraíba, Caixa Postal 5008, João Pessoa 58051-970, Paraíba, Brazil*

[†]*Programa de Pós-graduação em Física, Universidade Federal do Maranhão, Campus Universitário do Bacanga, São Luís 65080-805, Maranhão, Brazil*

[‡]*Coordenação do Curso de Física – Bacharelado, Universidade Federal do Maranhão, Campus Universitário do Bacanga, São Luís 65080-805, Maranhão, Brazil*

[§]*Departamento de Ciências Exatas e Educação, Instituto de Ciências, Agrárias, Exatas e Biológicas de Iturama – ICAEBI, Universidade Federal do Triângulo Mineiro – UFTM, Iturama, Minas Gerais 38280-000, Brazil*

In this work, we study the effects of vacuum Cherenkov radiation caused by nonminimal dimension-5 Lorentz-violating (LV) operators in the fermion sector. Explicitly, we focus on two independent isotropic pieces of each set of nonminimal coefficients. Under the assumption that vacuum Cherenkov radiation is an expected phenomenon, experimental data of ultra-high-energy cosmic rays (UHECRs) allow us to put stringent bounds on isotropic coefficients in quarks.

1. Introduction

Lorentz and CPT symmetries play a fundamental role in local relativistic field theories. If these symmetries are broken at the Planck scale, a multitude of effects can arise in the low-energy effective description via field theories. This concerns particle phenomenology, on the one hand, and gravitational waves, on the other hand. Prominent phenomena are shifts of energy levels of electrons in bound systems, anisotropies in particle and wave propagation, vacuum birefringence, and unusual processes such as Cherenkov-type radiation *in vacuo*. As a consequence, precise measurements of particle and gravitational-wave propagation can constrain several sets of Lorentz- and CPT-violating coefficients, testing to which extent these symmetries are exact in nature.

Cherenkov radiation occurs when a massive, charged particle travels through a medium with a velocity that exceeds the speed of light in this medium. After all, the latter is reduced by the refractive index of the material. At the same time, Lorentz symmetry breaking implies a nontrivial refractive index of the vacuum such that Cherenkov-type radiation *in vacuo* can be expected under certain circumstances. Indeed, various studies of vacuum Cherenkov radiation within the Standard-Model Extension (SME) have been carried out to date. Most of these analyses are based on modifications of the photon sector in the minimal SME, such as the Carroll–Field–Jackiw term[1,2] or CPT-even modified Maxwell theory.[3] Vacuum Cherenkov radiation has also been investigated in the gravity sector,[4] for pions,[5] partons,[6] and quarks.[7] These works have in common that they delve into the minimal SME, whereas a comprehensive analysis of nonminimal operators has not been carried out so far. In contrast to minimal coefficients, nonminimal ones are expected to become dominant at high energies. We intend to fill this gap partially by reporting on recent results on isotropic dimension-5 coefficients. Computational details will be given elsewhere.[8]

2. Basics

We explore modified Dirac fermions coupled to Maxwell electrodynamics, where the modification is governed by nonminimal LV operators of the SME fermion sector:

$$\mathcal{L}_{\mathrm{QED}} = \mathcal{L}_{\gamma} + \mathcal{L}_{\psi}, \tag{1a}$$

$$\mathcal{L}_{\gamma} = -\frac{1}{4}F^{\mu\nu}F_{\mu\nu} - \frac{1}{2}(\partial_{\mu}A^{\mu})^2, \tag{1b}$$

$$\mathcal{L}_{\psi} = \frac{1}{2}\bar{\psi}(\gamma^{\mu}i\mathcal{D}_{\mu} - m_{\psi} + \widehat{\mathcal{Q}})\psi + \text{H.c.}, \tag{1c}$$

with the $U(1)$ gauge field A_μ, the electromagnetic field strength tensor $F_{\mu\nu} = \partial_\mu A_\nu - \partial_\nu A_\mu$, the Dirac spinor field ψ, the Dirac conjugate field $\bar{\psi} := \psi^\dagger\gamma^0$, and the fermion mass m_ψ. All fields are defined in Minkowski spacetime with the metric tensor $\eta_{\mu\nu}$ of signature $(+,-,-,-)$. The Dirac matrices γ^μ satisfy the Clifford algebra $\{\gamma^\mu, \gamma^\nu\} = 2\eta^{\mu\nu}$. The photon sector is minimally coupled to the fermion sector via the gauge-covariant derivative $\mathcal{D}_\mu$. The operator $\widehat{\mathcal{Q}}$ contains all nonminimal LV coefficients to be considered as follows.

In the quantum regime, vacuum Cherenkov radiation is allowed when $\Delta E = 0$, where ΔE is the energy balance equation given as follows:

$$\Delta E = E(\mathbf{q}) - |\mathbf{k}| - E(\mathbf{q} - \mathbf{k}). \tag{2}$$

Here, $E(\mathbf{q})$ denotes the modified fermion dispersion relation, where $\mathbf{q}$ and $\mathbf{k}$ are the momenta of the initial fermion and the standard photon emitted, respectively. The range of the final-photon momentum is restricted by the solutions for the polar angle. Solving the energy balance equation leads to $\theta = \theta_0 = \theta_0(q, k, X_\subset)$, where $-1 \leq \cos\theta_0 \leq 1$, $q \equiv |\mathbf{q}|$, $k \equiv |\mathbf{k}|$, and $X_\subset$ is a subset of LV coefficients.

The decay rate Γ is calculated at tree level, where the corresponding Feynman diagram is presented in Fig. 1. The amplitude $\mathcal{M}$ reads

$$\mathcal{M} = e\bar{u}^{(s)}(q-k)\Gamma^\mu u^{(s')}(q)\epsilon_\mu^{(\lambda)}(k), \tag{3}$$

with the elementary charge e as well as the indices s and λ for spin and polarization states, respectively. Moreover, Γ^μ is a generalized Dirac matrix that includes information on nonminimal SME coefficients. For example, $\Gamma^\alpha = \gamma^\alpha - (a^{(5)})^{\mu\beta\alpha}\gamma_\mu p_\beta$ for the CPT-odd operator $\hat{a}^\mu$. Gauge invariance can be checked by computing $k_\mu\mathcal{M}^\mu = 0$ for $\mathcal{M}^\mu = e\bar{u}(q-k)\Gamma^\mu u(q)$.

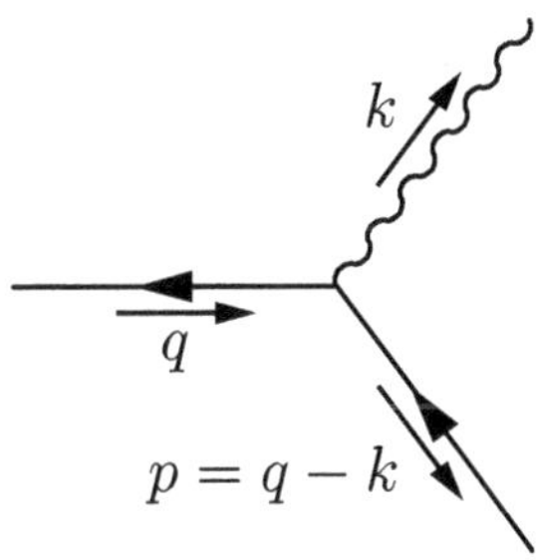

Fig. 1. Tree-level Feynman diagram for vacuum Cherenkov radiation. There is an incoming fermion with momentum $\mathbf{q}$ and the outgoing photon and fermion with momenta $\mathbf{k}$ and $\mathbf{p} = \mathbf{q} - \mathbf{k}$, respectively.

It is possible to show[7] that this decay rate for isotropic frameworks can be written as:

$$\Gamma = \frac{1}{2E(\mathbf{q})}\gamma, \tag{4a}$$

$$\gamma = \frac{1}{8\pi}\int_0^{k_{\max}} \mathrm{d}k\,\Pi(k)|\mathcal{M}|^2|_{\theta=\theta_0}, \tag{4b}$$

$$\Pi(k) = \frac{k\sin\theta}{E(\mathbf{q}-\mathbf{k})}\left|\frac{\partial\Delta E}{\partial\theta}\right|^{-1}\Bigg|_{\theta=\theta_0}, \tag{4c}$$

where $\Pi(k)$ is the phase space factor, and $|\mathcal{M}|^2$ is the amplitude squared averaged over the initial-fermion spin states and summed over the final-particle spin and polarization states. Moreover, $k_{\max}$ is the maximum possible value for the photon momentum, which is restricted by the requirement that $-1 \leq \cos\theta_0 \leq 1$.

3. Isotropic Sector of Dimension-5 Operator $\widehat{m}$

The nonminimal dimension-5 coefficients $m^{(5)}_{\alpha\beta}$ are CPT-even and possess two independent isotropic pieces, since the background tensor is not necessarily traceless. They are contained in the term

$$-\frac{1}{2}(m^{(5)})^{\alpha\beta}\bar{\psi}\mathrm{i}D_{(\alpha}\mathrm{i}D_{\beta)}\psi + \mathrm{H.c.} \subset \mathcal{L}^{(5)}_{\psi D}, \tag{5}$$

of Tab. I in Ref. 9. The nonminimal LV operator within the modified Dirac equation in momentum space is given by[10]

$$\widehat{\mathcal{Q}}^{(5)} = -m^{(5)}_{\alpha\beta}p^\alpha p^\beta \supset -\mathring{m}^{(5)}_0(p^0)^2 - \mathring{m}^{(5)}_2 p^2, \tag{6}$$

where $\mathbf{p}^2 \equiv p^2$, $\mathring{m}^{(5)}_0 \equiv m^{(5)}_{00}$, and $\mathring{m}^{(5)}_2 \equiv m^{(5)}_{11} = m^{(5)}_{22} = m^{(5)}_{33}$ are the isotropic parts. We will drop the superscript '(5)', for brevity.

To find the dispersion relations, we look for nontrivial solutions of the modified Dirac equation $(\not{p} - m_\psi + \widehat{\mathcal{Q}})\psi = 0$ and solve the determinant equation for p_0. First, there are four spurious dispersion relations:

$$\begin{aligned} E &= \pm\frac{1}{\sqrt{2}}\sqrt{\frac{1}{\mathring{m}_0^2} - \frac{2m_\psi}{\mathring{m}_0} + \frac{\sqrt{1-4m_\psi\mathring{m}_0 - 4\mathring{m}_0^2p^2}}{\mathring{m}_0^2}} \\ &= \pm\frac{1}{\mathring{m}_0} \mp m_\psi \mp (p^2 + 2m_\psi^2)\frac{\mathring{m}_0}{2} + \cdots . \end{aligned} \tag{7}$$

In the Lorentz-invariant limit, these expressions do not reproduce the conventional dispersion relations for massive particles. Instead, they are singular. In principle, such spurious dispersion relations can be interpreted as describing Planck-scale effects, whereupon their meaning in effective field theory can be questioned. For this reason, and since our analysis is restricted to tree-level effects, Eq. (7) will be ignored.

The remaining four dispersion relations are perturbatively consistent and spin-degenerate. The positive-energy dispersion relation, which holds for particles, reads

$$\begin{aligned} E_{\mathring{m}_0} &= \frac{1}{\sqrt{2}}\sqrt{\frac{1}{\mathring{m}_0^2} - \frac{2m_\psi}{\mathring{m}_0} - \frac{\sqrt{1-4m_\psi\mathring{m}_0 - 4\mathring{m}_0^2p^2}}{\mathring{m}_0^2}} \\ &= (m_\psi\mathring{m}_0 + 1)\sqrt{p^2+m_\psi^2} + \cdots . \end{aligned} \tag{8}$$

Equation (8) enables us to solve the energy balance equation (2) for the polar angle θ. Restricting ourselves to first order in the isotropic coefficient and employing $|\mathbf{q}| \equiv q$, we find

$$\cos\theta = \frac{1}{q}\left[\sqrt{q^2+m_\psi^2} + m_\psi\mathring{m}_0\left(k - \sqrt{q^2+m_\psi^2}\right)\right]. \tag{9}$$

The magnitude of the photon momentum has a maximum corresponding to the maximum value of $\cos\theta$. This maximum momentum to first order in $\mathring{m}_0$ is given by:

$$\begin{aligned} k_{\max} &= \frac{q-\sqrt{q^2+m_\psi^2}}{m_\psi\mathring{m}_0} + \frac{1}{4}m_\psi\mathring{m}_0\left(\sqrt{q^2+m_\psi^2}+q\right) \\ &\quad + \frac{1}{2}\left(3q - \sqrt{q^2+m_\psi^2}\right). \end{aligned} \tag{10}$$

Vacuum Cherenkov radiation takes place for positive k, where a minimum nonzero initial-fermion energy follows from the relation $k_{\max} = 0$. Thus, the process occurs when the initial-fermion energy lies above a certain threshold given by

$$q_{\text{th}} = \sqrt{\frac{m_\psi}{3\mathring{m}_0}} + \cdots , \tag{11}$$

where we see that only positive values of the isotropic coefficient $\mathring{m}_0$ are permitted. It is observed that $q_{\text{th}} \to \infty$ for $\mathring{m}_0 \to 0$, as expected, since the process is forbidden in the Lorentz-invariant limit. The phase space factor follows from the kinematics according to Eq. (4c). In addition to the kinematic ingredients, we also need to take into account the dynamics of the process, which is described by the amplitude squared. Details on how to compute the latter will be provided in Ref. 8.

The same procedure can be repeated for the other independent isotropic piece governed by $\mathring{m}_2$. Since the latter is not accompanied by time derivatives, spurious dispersion relations do not emerge. The perturbative positive-energy dispersion relation reads

$$E_{\mathring{m}_2} = \sqrt{m_\psi^2 + (1 + 2m_\psi \mathring{m}_2)p^2 + \mathring{m}_2^2 p^4}. \tag{12}$$

In this case, the polar angle and the maximum energy of the photon in terms of q are complicated functions even at leading order in $\mathring{m}_2$. Nevertheless, the procedure for finding the threshold energy is the same as before. Doing so provides a result similar to Eq. (11):

$$q_{\text{th}} = \sqrt{\frac{m_\psi}{3\mathring{m}_2}} + \cdots. \tag{13}$$

The coefficient $\mathring{m}_2$ must also be positive for the process to occur and the threshold is singular in $\mathring{m}_2$, as expected.

The decay rates for each of the two settings with either nonzero $\mathring{m}_0$ or $\mathring{m}_2$ are presented in Fig. 2. In fact, both curves cannot be distinguished

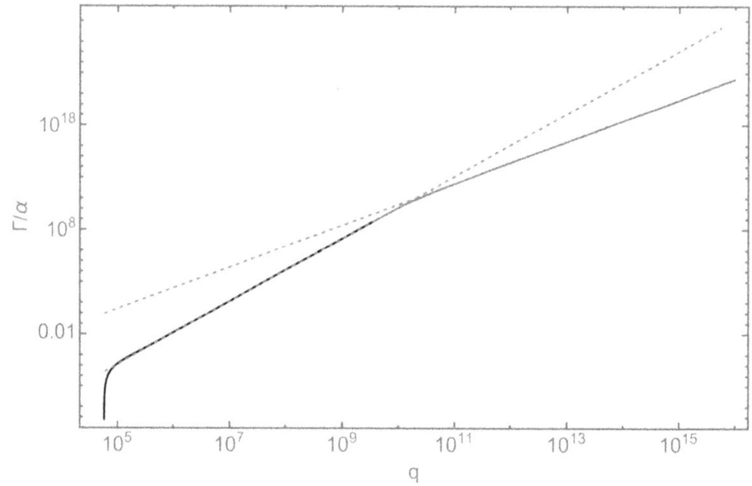

Fig. 2. Double-logarithmic plot of the decay rates Γ/α with the fine-structure constant $\alpha = e^2/(4\pi)$ as functions of the initial-fermion momentum q for $m_\psi = 1.00\,\text{GeV}$ as well as the isotropic coefficients $\mathring{m}_0 = 10^{-10}\,\text{GeV}^{-1}$ and $\mathring{m}_2 = 10^{-10}\,\text{GeV}^{-1}$, respectively. Both curves lie on top of each other.

from each other. Several comments are in order. First, the equal threshold momenta can clearly be perceived, since $\Gamma \to 0$ when $q \to q_{\text{th}}$, as expected. Second, the curve for $\mathring{m}_2$ possesses a feature resembling a knee, where the behavior of the decay rate changes drastically. Interestingly, a comparable behavior was observed in Ref. 2 for the vacuum Cherenkov process with modified photons. Third, the curve for $\mathring{m}_0$ stops before this knee, since Eq. (8) becomes complex for momenta exceeding a certain upper limit.

4. Isotropic Sector of Dimension-5 Operator $\hat{a}^{\mu}$

The nonminimal dimension-5 coefficients $a^{(5)}_{\alpha\beta\mu}$ are CPT-odd and have two independent isotropic pieces as does $m^{(5)}_{\mu\nu}$ studied previously. They are contained in the term

$$-\frac{1}{2}(a^{(5)})^{\mu\alpha\beta}\bar{\psi}\gamma_{\mu}\mathrm{i}D_{(\alpha}\mathrm{i}D_{\beta)}\psi + \text{H.c.} \subset \mathcal{L}^{(5)}_{\psi D}, \tag{14}$$

of Tab. I in Ref. 9. The operator in the modified Dirac equation reads $\widehat{\mathcal{Q}} = -(a^{(5)})^{\mu\alpha\beta}p_{\alpha}p_{\beta}\gamma_{\mu}$ such that the isotropic parts are formulated as follows[10]:

$$(a^{(5)})^{\mu\alpha\beta}p_{\alpha}p_{\beta}\gamma_{\mu} \supset \mathring{a}^{(5)}_0 p^0p^0\gamma^0 + \frac{1}{3}\mathring{a}^{(5)}_2(p^kp^k\gamma^0 + 2p^0p^k\gamma^k), \tag{15}$$

where the superscript '(5)' will be omitted, for brevity. The dispersion relations follow again from solving the modified Dirac equation in momentum space. For the isotropic part $\mathring{a}_0$, the condition of a vanishing determinant of the Dirac operator implies four spurious dispersion relations, such as for $\mathring{m}_0$. The perturbative dispersion relations are again spin-degenerate. The positive-energy result at first order in the isotropic coefficient is

$$E_{\mathring{a}_0} = \sqrt{p^2 + m_{\psi}^2} + \mathring{a}_0\left(p^2 + m_{\psi}^2\right) + \cdots. \tag{16}$$

By solving the energy balance equation (2) with Eq. (16), we find the polar angle. As previously, the maximum photon momentum $k_{\max}$ follows from the maximum of $\cos\theta$. The threshold momentum of the initial fermion is computed from $k_{\max} = 0$. Performing a computation exact in $\mathring{a}_0$ leads to

$$q_{\text{th}} = \sqrt[3]{\frac{m_{\psi}^2}{4\mathring{a}_0}} + \cdots, \tag{17}$$

where $q_{\text{th}} \to \infty$ holds for $\mathring{a}_0 \to 0$, as expected.

Finally, we proceed to study the effects caused by the isotropic coefficient $\mathring{a}_2$ of Eq. (15). Since the latter is not contracted with time derivatives,

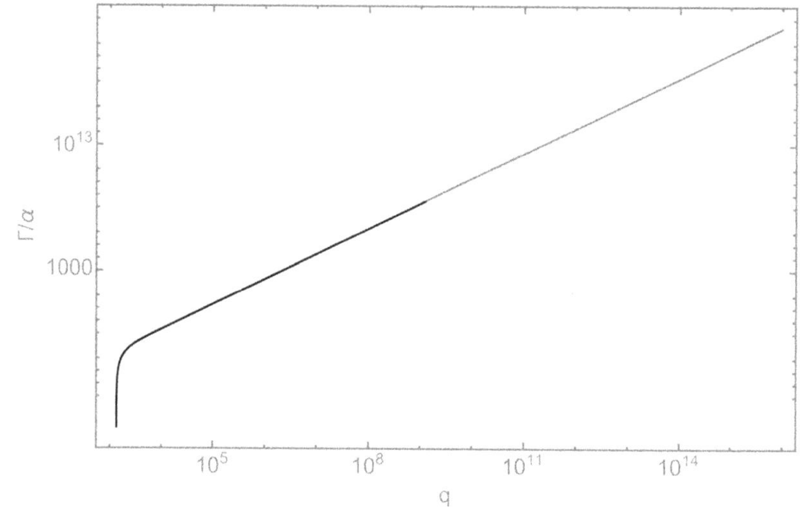

Fig. 3. The same as Fig. 2 for the isotropic coefficients $\mathring{a}_0 = 10^{-10}\,\mathrm{GeV}^{-1}$ and $\mathring{a}_2 = 10^{-10}\,\mathrm{GeV}^{-1}$, respectively. One curve lies on top of the other.

there are no spurious dispersion relations. The positive-energy perturbative one reads

$$E_{\mathring{a}_2} = \sqrt{m_\psi^2 + p^2} + \mathring{a}_2 p^2 + \cdots . \tag{18}$$

Comparing Eqs. (16) and (18), we see that the former differs from the latter only by an additional fermion mass term. The threshold momentum has a behavior analogous to that of Eq. (17):

$$q_{\mathrm{th}} = \sqrt[3]{\frac{m_\psi^2}{4\mathring{a}_2}} + \cdots . \tag{19}$$

Let us look at the decay rates for $\mathring{a}_0$ and $\mathring{a}_2$, which are presented in Fig. 3. The curves are indistinguishable. In contrast to the decay rate for $\mathring{m}_0$ in Fig. 2, the curve for $\mathring{a}_2$ does not exhibit a knee. Note that the curve for $\mathring{a}_0$ again stops at a certain maximum momentum, where the exact expression for the energy becomes complex.

5. Constraints on the Isotropic Coefficients

If we assume that vacuum Cherenkov radiation is an expected phenomenon, it is possible to use data collected from UHECRs to put stringent bounds on the previous nonminimal, isotropic LV coefficients for the quark sector.

To do so, we employ an effective description based on the modified QED introduced in Eq. (1) and put the electroweak and strong interactions aside.

A primary cosmic ray observed on Earth has energy $E < E_{\text{th}}$, where E_{th} is the fermion threshold energy following from the threshold momentum q_{th} of the previous sections. The latter can be expressed in generic form as $q_{\text{th}} = \rho m_\psi^\lambda / \mathring{X}^\sigma$, where ρ, λ and σ are dimensionless parameters and $\mathring{X}$ is an isotropic coefficient. The inequality $E < E_{\text{th}}$ leads us to

$$\mathring{X} < \left(\frac{\rho m_\psi^\lambda}{E}\right)^{\frac{1}{\sigma}} . \tag{20}$$

Twice the experimental energy uncertainty ΔE is added to arrive at bounds with 2σ confidence level (CL):

$$\mathring{X} < \left(\frac{\rho m_\psi^\lambda}{E}\right)^{\frac{1}{\sigma}} + 2\Delta E \left| \frac{\partial}{\partial E} \left(\frac{\rho m_\psi^\lambda}{E}\right)^{\frac{1}{\sigma}} \right| . \tag{21}$$

We consider measurements reported by the Pierre-Auger observatory. In particular, the event 737165 from Tab. I in Ref. 11 is consulted, whose energy measured is $E = 212 \times 10^{18}$ eV with an uncertainty of 25%. The composition of the cosmic-ray primary is unknown, but we repeat the analysis done for the minimal-SME coefficients,[7] where an iron nucleus is assumed to have been the primary. The energy of a single nucleon is E/N with $N = 56$ being the number of nucleons in the iron nucleus.

In addition, we assume that a Cherenkov photon is emitted by an up- or a down-quark of a nucleon. This quark carries a fraction $r = 0.1$ of the nucleon energy and the quark masses considered are $m_{\text{u}} \approx 2.16 \times 10^{-3}$ GeV and $m_{\text{d}} \approx 4.70 \times 10^{-3}$ GeV. The results of the constraints on the isotropic

Table 1. Constraints on isotropic coefficients in the u- and d-quark sectors at 2σ CL based on the event 737165 of Ref. 11.

Fermion sector	Isotropic coefficient	Upper constraint
u quark	$\mathring{m}_0$	$< 3 \times 10^{-18}\,\text{GeV}^{-1}$
	$\mathring{m}_2$	$< 3 \times 10^{-18}\,\text{GeV}^{-1}$
	$\mathring{a}_0$	$< 2 \times 10^{-29}\,\text{GeV}^{-1}$
	$\mathring{a}_2$	$< 2 \times 10^{-29}\,\text{GeV}^{-1}$
d quark	$\mathring{m}_0$	$< 6 \times 10^{-18}\,\text{GeV}^{-1}$
	$\mathring{m}_2$	$< 6 \times 10^{-18}\,\text{GeV}^{-1}$
	$\mathring{a}_0$	$< 9 \times 10^{-29}\,\text{GeV}^{-1}$
	$\mathring{a}_2$	$< 9 \times 10^{-29}\,\text{GeV}^{-1}$

coefficients are presented in Table 1. These bounds are better by many orders of magnitude than the ones obtained for some of the minimal coefficients in quarks.[7] This is to be expected, since the effects of nonminimal coefficients increase with energy as compared to those of the minimal coefficients.

References

1. R. Lehnert and R. Potting, *Phys. Rev. D* **70**, 125010 (2004); R. Lehnert and R. Potting, *Phys. Rev. Lett.* **93**, 110402 (2004); C. Kaufhold and F.R. Klinkhamer, *Nucl. Phys. B* **734**, 1 (2006); C. Kaufhold and F.R. Klinkhamer, *Phys. Rev. D* **76**, 025024 (2007); M.A. Hohensee, R. Lehnert, D.F. Phillips, and R.L. Walsworth, *Phys. Rev. D* **80**, 036010 (2009); C. Adam and F.R. Klinkhamer, *Nucl. Phys. B* **607**, 247 (2001); B. Altschul, *Phys. Rev. D* **90**, 021701(R) (2014); K. Schober and B. Altschul, *Phys. Rev. D* **92**, 125016 (2015); D. Colladay, P. McDonald, J.P. Noordmans, and R. Potting, *Phys. Rev. D* **95**, 025025 (2017).
2. D. Colladay, P. McDonald, and R. Potting, *Phys. Rev. D* **93**, 125007 (2016).
3. B. Altschul, *Phys. Rev. Lett.* **98**, 041603 (2007); F.R. Klinkhamer and M. Schreck, *Phys. Rev. D* **78**, 085026 (2008).
4. M. Schreck, *Symmetry* **10**, 424 (2018); V.A. Kostelecký and J.D. Tasson, *Phys. Lett. B* **749**, 551 (2015).
5. B. Altschul, *Phys. Rev. D* **93**, 105007 (2016).
6. J.S. Díaz and F.R. Klinkhamer, *Phys. Rev. D* **92**, 025007 (2015).
7. M. Schreck, *Phys. Rev. D* **96**, 095026 (2017).
8. A.Yu. Petrov, M. Schreck, and A.R. Vieira, *Eur. Phys. J. C* **86**, 30 (2026).
9. V.A. Kostelecký and Z. Li, *Phys. Rev. D* **99**, 056016 (2019).
10. V.A. Kostelecký and M. Mewes, *Phys. Rev. D* **88**, 096006 (2013).
11. J. Abraham *et al.* (Pierre Auger Collaboration), *Astropart. Phys.* **27**, 155 (2007).

Chapter 20

Neutron EDM Experiment with an Advanced Ultracold Neutron Source at TRIUMF

T. Higuchi,[*,†] B. Algohi,[‡] D. Anthony,[§] L. Barrón-Palos,[¶]
M. Bradley,[‖] A. Brossard,[§] T. Bui,[‡] J. Chak,[§] R. Chiba,[§,**]
C. Davis,[§] R. de Vries,[††] K. Drury,[§] D. Fujimoto,[§] R. Fujitani,[*,‡‡]
M. Gericke,[‡] P. Giampa,[§] R. Golub,[§§] T. Hepworth,[††]
G. Ichikawa,[¶¶] S. Imajo,[†] A. Jaison,[‡] B. Jamieson,[††]
M. Katotoka,[††] S. Kawasaki,[‖‖] M. Kitaguchi,[***] W. Klassen,[†††]
E. Korkmaz,[‡‡‡] E. Korobkina,[§§] M. Lavvaf,[‡] T. Lindner,[§,††]
N. Lo,[§] S. Longo,[‡] K. Madison,[†††] Y. Makida,[‖‖] J. Malcolm,[§]
J. Mammei,[‡] R. Mammei,[††] Z. Mao,[†††] C. Marshall,[§]
J.W. Martin,[††] M. McCrea,[††] E. Miller,[†††] M. Miller,[§§§]
K. Mishima,[†] T. Mohammadi,[‡] T. Momose,[†††] T. Okamura,[‖‖]
H.J. Ong,[†] R. Patni,[§] R. Picker,[§,**] K. Qiao,[†,¶¶¶]
W.D. Ramsay,[§] W. Rathnakela,[‡] D. Salazar,[§,**] J. Sato,[‖‖]
W. Schreyer,[§,‖‖‖] T. Shima,[†] H.M. Shimizu,[‖‖] S. Sidhu,[§]
S. Stargardter,[‡] P. Switzer,[††] I. Tanihata,[†] Tushar,[‡]
S. Vanbergen,[§] W.T.H. van Oers,[‡,§] Y. Watanabe,[‖‖]
N. Yazdandoost,[§] Q. Ye,[†††] A. Zahra,[‡] and M. Zhao[§]

(TUCAN Collaboration)

[*] *Institute for Integrated Radiation and Nuclear Science, Kyoto University, Kumatori, 590-0494, Japan*

[†] *Research Center for Nuclear Physics, the University of Osaka, Ibaraki, 567-0047, Japan*

[‡] *The University of Manitoba, Winnipeg, MB R3T 2N2, Canada*

[§] *TRIUMF, Vancouver, BC V6T 2A3, Canada*

[¶] *Instituto de Física, Universidad Nacional Autónoma de México, Apartado Postal 20-364, 01000, Mexico*

[‖]*The University of Saskatchewan, Saskatoon, SK S7N 5A2, Canada*
[**]*Simon Fraser University, Burnaby, BC V5A 1S6, Canada*
[††]*The University of Winnipeg, Winnipeg, MB R3B 2E9, Canada*
[‡‡]*Department of Nuclear Engineering, Kyoto University, Kyoto, 615-8540, Japan*
[§§]*North Carolina State University, Raleigh, NC 27695-8202, USA*
[¶¶]*Institute of Materials Structure Science, High Energy Accelerator Research Organization (KEK), Tokai, 319-1106, Japan*
[‖‖]*Institute for Particle and Nuclear Studies, High Energy Accelerator Research Organization (KEK), Tsukuba, 305-0801, Japan*
[***]*Nagoya University, Nagoya, 464-8601, Japan*
[†††]*The University of British Columbia, Vancouver, BC V6T 1Z4, Canada*
[‡‡‡]*The University of Northern British Columbia, Prince George, BC V2N 4Z9, Canada*
[§§§]*McGill University, Montréal, QC H3A 0G4, Canada*
[¶¶¶]*Graduate School of Science, The University of Osaka, Toyonaka, 560-0043, Japan*
[‖‖‖]*Oak Ridge National Laboratory, Knoxville, TN 37830, USA*

The TRIUMF Ultracold Advanced Neutron (TUCAN) collaboration has been developing a high-intensity ultracold neutron (UCN) source aimed at searching for the neutron electric dipole moment (EDM) with a sensitivity goal of $10^{-27}\,e\,\mathrm{cm}$. This article reports on recent progress in the commissioning of the UCN source and the development of the neutron EDM spectrometer. In its final configuration, the accelerator-driven super-thermal UCN source will enable a neutron EDM experiment with two orders of magnitude improved statistics compared to the current best experiment. Substantial progress in 2024 allowed the collaboration to operate the complete source system, with the exception of the liquid deuterium cold moderator, resulting in the first production of UCNs. The status of the EDM spectrometer is also presented, with emphasis on UCN handling components and magnetic subsystems relevant to field control, shielding, and magnetometry.

1. Background

Measurements of the neutron electric dipole moment (EDM) occupy an important place in today's particle physics. A finite neutron EDM would violate time-reversal symmetry, which, on the basis of CPT symmetry, would imply CP violation. Experiments so far have found results consistent with zero, placing stringent constraints on the QCD $\bar{\theta}$ term and on theories beyond the Standard Model, such as supersymmetric extensions or multi-Higgs-doublet models.[1–5]

State-of-the-art neutron EDM experiments utilize stored ultracold neutrons (UCNs) to measure the EDM by observing the precession of the neutron spin under electric and magnetic fields. UCNs are neutrons with kinetic energies of around 300 neV or less, which can be stored in containers coated with materials of high Fermi potential. This enables neutron EDM experiments with reduced systematic effects related to neutron velocity and with observation times on the order of 100 s. The current best upper limit, obtained at the Paul Scherrer Institute, is $|d_n| < 1.8 \times 10^{-26}\, e\text{cm}$ (90% C.L.).[6] This result is limited by the statistical uncertainty arising from the number of UCNs available during the experiment.

To improve upon this limit and measure the neutron EDM with unprecedented precision, the TRIUMF Ultracold Advanced Neutron (TUCAN) collaboration has been developing a high-intensity UCN source, expected to increase the number of usable UCNs by two orders of magnitude. Our goal sensitivity is $10^{-27}\, e\text{cm}$ and is competitive with other next-generation EDM experiments that are in preparation in locations around the world.[7–9]

2. Advanced Ultracold Neutron Source at TRIUMF

The TUCAN source is based on a combination of accelerator-driven spallation neutron production and the super-thermal method with superfluid helium (He-II) as the UCN converter.[10] High-energy neutrons produced by spallation are decelerated by cold moderators and reach the He-II converter, where they are downscattered to become UCNs. This scheme was developed at Research Center for Nuclear Physics in Japan and has been demonstrated with a prototype UCN source.[11] The prototype source was transported to TRIUMF, installed on a dedicated 480-MeV proton beamline driven by the TRIUMF cyclotron,[12] and was used to produce the first UCNs at TRIUMF in 2017.[13] It was operated until 2019 for characterization of the source itself and tests of UCN handling components.[13–15]

Based on experience with the prototype source, the collaboration has been developing a new, upgraded UCN source featuring a liquid deuterium (LD_2) cold moderator optimized for UCN production, a 27-L He-II converter, and a helium cryostat capable of maintaining He-II temperatures near 1 K under the expected 10 W heat load during beam irradiation.[16,18] The expected UCN yield with a 40 μA proton beam is 1.4×10^7 UCN/s,[16] representing a 500-fold increase over the 2×10^4 UCN/s achieved with the prototype source operated at the nominal 1 μA beam current.[13]

According to Monte-Carlo UCN transport simulations, the source will enable an EDM experiment with 10^6 UCNs detected per measurement cycle. This corresponds to a statistical uncertainty of $10^{-25}\,e$cm per cycle and will allow us to reach the target sensitivity of $10^{-27}\,e$cm within 280 days of data taking.[19]

3. Recent Status

Status of the UCN source The year 2024 saw major progress in the manufacturing of components for the TUCAN source. In April, fabrication of the vessels for the UCN production volume was completed. This was followed by the completion of key components, including the main heat exchanger of the helium cryostat and the UCN guides for extracting the produced UCNs from the source. By the end of October, all hardware for the TUCAN source, except the LD_2 system, was in place. The helium cryostat, built in Japan and shipped to TRIUMF in 2021, has undergone a series of cooldowns to test its performance, initially using natural helium and later using dedicated helium isotopes: ^{3}He for the cryostat cooling line and isotopically pure ^{4}He for the UCN converter.

In November 2024, the full TUCAN source system apart from the LD_2 moderator was commissioned. The UCN production volume was filled with isotopically pure ^{4}He cooled to superfluid temperatures. Although the absence of the LD_2 moderator was expected to significantly reduce the yield, a production rate of $\sim 10^4$ UCN/μA was still anticipated. However, no significant UCN events above background were observed. Air contamination in the isotopically pure ^{4}He batch used in the production volume was strongly suspected as the cause, supported by several observations made before and during the run. Nevertheless, the cryogenic performance of the source was successfully characterized using both a heater and the proton beam, demonstrating sufficient cooling capacity for operation at 40 μA.[20] Cold neutron fluxes from the D_2O moderator were measured by the gold foil activation method to benchmark Monte Carlo N-Particle (MCNP) simulations used in the source design.[16]

After a large amount of contaminants had been removed from the ^{4}He, another attempt was made in June 2025 with a custom-built ^{3}He/^{4}He purification system installed on the isotopically pure ^{4}He inlet. This led to the first detection of UCNs extracted from the TUCAN source. The initial results indicate good agreement between the estimated and observed UCN yields.[17]

Development of the EDM spectrometer: In parallel with the development of the UCN source, subsystems of the neutron EDM spectrometer have been advanced. Tests of key UCN handling components have been conducted using the pulsed UCN source at J-PARC/MLF,[21] which has supported various experiments, including transmission through UCN guides, storage in a prototype EDM measurement cell, and UCN polarization using magnetized iron films.[22–24] The cold-neutron beamline MINE2 at JRR-3 has also been used for complementary characterization of polarization films and surface coatings relevant to UCN applications.[24,25]

One crucial aspect of the neutron EDM experiment is the precise control of magnetic fields, an area in which active development is ongoing. In the final neutron EDM experiment, the effective stability required for the spin-holding magnetic field ($B_0 \approx 1\ \mu$T) is about 10 fT over averaging times on the order of 100 s. This is achieved through the use of advanced magnetic shielding and an optically pumped ^{199}Hg magnetometer that monitors magnetic field variations in the same volume occupied by UCNs.

A multilayer magnetically shielded room (MSR), consisting of five mu-metal layers and one copper layer, has recently been constructed. It has an outer dimension of 3.5 m and an inner room measuring 2.25 m on each side. The shielding performance was characterized at each stage of assembly, demonstrating shielding sufficient to achieve field stability at the 10 pT level. During the tests, ambient magnetic fields from the TRIUMF cyclotron, reaching up to 350 μT around the MSR, were found to degrade the shielding performance. Compensation coils have been designed to address this issue and are currently under construction.[26]

The ^{199}Hg co-magnetometer under development at the University of British Columbia has demonstrated a sensitivity on the order of 1 pT, currently limited by a few aspects of the prototype setup, such as magnetic field inhomogeneities and wall collisions in the small test cell. Once integrated into the full-size EDM measurement cell in the MSR, it is expected to reach the sensitivity required to monitor magnetic field variations at the 10 fT level.

Generation of a uniform B_0 field is also important, since magnetic-field inhomogeneities are one of the dominant sources of systematic uncertainty.[6,27] To map magnetic field distributions inside the MSR, an array of optically pumped Cs atomic magnetometers have been developed.

Recent experiments in a test setup demonstrated a stability of 90 fT over an averaging time of 150 s.[28] The system supports operation of multiple sensors using a single laser diode; it has been successfully tested with 3 sensors and is being prepared for deployment with 20.[28] Coil systems with different purposes, including a self-shielded B_0 coil and a set of shim coils to correct for magnetic field inhomogeneities, have been designed and are currently under construction.[29]

4. Potential Tests of Lorentz Symmetry with TUCAN

The advanced magnetic subsystems developed for the neutron EDM experiment can, in principle, be applied to clock-comparison tests of Lorentz symmetry, in which Lorentz-violating effects would manifest as sidereal modulations in the spin-precession frequencies of magnetometers. Here, the use of multiple spin species is essential to cancel background magnetic-field drifts and isolate components that would couple differently to each. Physical systems accessible with TUCAN include neutron/^{199}Hg, Cs/^{199}Hg, and ^{199}Hg/^{201}Hg. Since the Cs magnetometer measures the precession of an electron spin and the Hg magnetometer measures that of a nuclear spin,

Table 1. List of minimal SME coefficients accessible to the TUCAN EDM spectrometer, excerpted from Ref. 39. For the limit from neutron/Hg systems, the result from Ref. 35 is taken, which updates the earlier result in Ref. 34.

Sector	Coefficients	Limits (GeV)	Systems	Refs.
e	r_e	$< 3.2 \times 10^{-24}$	Hg/Cs	[30]
	$\lvert\tilde{b}_J\rvert, (J = X, Y)$	$< 10^{-27}$	Hg/Cs	[31, 32]
	$\lvert\tilde{d}_J\rvert, (J = X, Y)$	$< 10^{-22}$	Hg/Cs	[31, 32]
	$\lvert\tilde{g}_{D,J}\rvert, (J = X, Y)$	$< 10^{-22}$	Hg/Cs	[31, 32]
p	$\tilde{b}_Z$	$< 7 \times 10^{-29}$	Hg/Cs	[33]
	$\tilde{b}_\perp$	$< 4 \times 10^{-30}$	Hg/Cs	[33]
	$\lvert\tilde{b}_J\rvert, (J = X, Y)$	$< 10^{-27}$	Hg/Cs	[31, 32]
	$\tilde{d}_J, \tilde{g}_{D,J}, (J = X, Y)$	$< 3 \times 10^{-28}$	Hg/Cs	[33]
	$\lvert\tilde{d}_J\rvert, \lvert\tilde{g}_{D,J}\rvert, (J = X, Y)$	$< 10^{-25}$	Hg/Cs	[31, 32]
n	$\tilde{b}_Z$	$< 7 \times 10^{-30}$	Hg/Cs	[33]
	$\tilde{b}_\perp$	$< 3.7 \times 10^{-31}$	Hg/Cs	[33]
	$b_\perp$	$< 1 \times 10^{-29}$	n/Hg	[35]
	r_n	$< 1.5 \times 10^{-30}$	Hg/Cs	[30]
	$\lvert\tilde{b}_J\rvert, (J = X, Y)$	$< 10^{-30}$	Hg/Cs	[31, 32]
	$\lvert\tilde{c}_-\rvert, \lvert\tilde{c}_Z\rvert$	$< 10^{-27}$	Hg/Hg, Ne/He	[36, 38]
	$\tilde{d}_J, \tilde{g}_{D,J}, (J = X, Y)$	$< 3 \times 10^{-29}$	Hg/Cs	[33]
	$\lvert\tilde{d}_J\rvert, \lvert\tilde{g}_{D,J}\rvert, (J = X, Y)$	$< 10^{-28}$	Hg/Cs	[31, 32]

the Cs/^{199}Hg system is sensitive to Lorentz-violating effects in the electron, proton, and neutron sectors.[30,31,33] In contrast, the neutron/^{199}Hg and ^{199}Hg/^{201}Hg systems are primarily sensitive to the neutron sector.[35–37] Constraints on minimal Standard-Model Extension (SME) coefficients obtained from previous experiments using such systems are excerpted from the data tables and summarized in Table 1.[39] A recently developed theoretical treatment further allows constraints to be placed on nonminimal Lorentz-violating coefficients from these experiments.[40]

Some clock-comparison experiments, including that in Ref. 33, employed rotating apparatus to enhance sensitivity. While this approach is not feasible with the TUCAN MSR, the projected effective magnetic field stability of 10–100 fT per 100-s cycle makes it possible to improve the limits on these coefficients. For reference, upper limits on Lorentz-violating signals in units of magnetic fields were 14 fT in Ref. 31 and 4 fT in Ref. 33.

5. Summary and Outlook

In this contribution, we report on the recent commissioning progress of the TUCAN source at TRIUMF. We also outline the development status of the EDM spectrometer, with a focus on UCN handling components and the magnetic subsystems essential to the EDM experiment.

The major milestone of the first UCN production from the TUCAN source was achieved in June 2025. The next significant step is the commissioning of the full source, including the LD_2 moderator. The LD_2 cryostat was completed in April 2025 and delivered to TRIUMF in May 2025. Integration of the LD_2 moderator is expected to enhance the UCN production rate by a factor of 30, unlocking the full performance of the TUCAN source.

In 2026, a site-wide shutdown of the TRIUMF accelerators is scheduled. The TUCAN collaboration will use this period to characterize the performance of spectrometer subsystems and commission the EDM spectrometer in preparation for UCN experiments beginning in 2027.

Acknowledgments

We gratefully acknowledge the support of the Canada Foundation for Innovation; the Canada Research Chairs program; the Natural Sciences and Engineering Research Council of Canada (NSERC) SAPPJ-2016-00024, SAPPJ-2019-00031, and SAPPJ-2023-00029; JSPS KAKENHI (Grant Nos. 18H05230, 19K23442, 20KK0069, 20K14487, 22H01236, 25H00652 and

25H02182); JSPS Bilateral Program (Grant No. JSPSBP120239940); JST FOREST Program (Grant No. JPMJFR2237); International Joint Research Promotion Program of Osaka University; RCNP COREnet; the Yamada Science Foundation; the Murata Science Foundation; Yazaki Memorial Foundation for Science and Technology; the Grant for Overseas Research by the Division of Graduate Studies (DoGS) of Kyoto University; the United States Department of Energy (DoE) (grant No. DE-FG02-97ER41042); the Universidad Nacional Autónoma de México–DGAPA program PASPA and grant PAPIIT AG102023.

References

1. M. Pospelov and A. Ritz, *Ann. Phys.* **318**, 119 (2005).
2. J. Engel, M.J. Ramsey-Musolf, and U. Van Kolck, *Prog. Part. Nucl. Phys.* **71**, 21 (2013).
3. J. Liang *et al.* (χQCD collaboration), *Phys. Rev. D* **108**, 094512 (2023).
4. W.-S. Hou, G. Kumar, and S. Teunissen, *Phys. Rev. D* **109**, L011703 (2024).
5. T. Abe *et al.*, *Phys. Rev. D* **98**, 075029 (2018).
6. C. Abel *et al.*, *Phys. Rev. Lett.* **124**, 081803 (2020).
7. D. Wurm *et al.*, *EPJ Web Conf.* **219**, 02006 (2019).
8. N.J. Ayres *et al.*, *Eur. Phys. J. C* **81**, 512 (2021).
9. D.K.-T. Wong *et al.*, *Nucl. Inst. Meth. A* **1050**, 168105 (2023).
10. R. Golub and S.K. Lamoreaux, *Phys. Lett. A* **62**, 338 (1977).
11. Y. Masuda *et al.*, *Phys. Rev. Lett.* **108**, 134801 (2012).
12. S. Ahmed *et al.*, *Nucl. Inst. Meth. A* **927**, 101 (2019).
13. S. Ahmed *et al.* (TUCAN Collaboration), *Phys. Rev. C* **99**, 025503 (2019).
14. S. Hansen-Romu, Ph.D. thesis, the University of Manitoba (2023).
15. H. Akatsuka *et al.*, *Nucl. Inst. Meth. A* **1049**, 168106 (2023).
16. W. Schreyer *et al.*, *Nucl. Inst. Meth. A* **959**, 163525 (2020).
17. B. Algohi *et al.*, arXiv:2509.02916.
18. S. Kawasaki, T. Okamura, and the TUCAN Collaboration, *IOP Conf. Ser.: Mater. Sci. Eng.* **755**, 012140 (2020).
19. S. Sidhu *et al.*, *EPJ Web Conf.* **282**, 01015 (2023).
20. J.W. Martin *et al.*, *EPJ Web Conf.* **333**, 03004 (2025).
21. S. Imajo *et al.*, *Prog. Theor. Exp. Phys.* **2016**, 013C02 (2016).
22. S. Imajo *et al.*, *Phys. Rev. C* **108**, 034605 (2023).
23. S. Vanbergen, Ph.D. thesis, the University of British Columbia (2025).
24. T. Higuchi *et al.*, *J. Phys. Soc. Jpn.* **93**, 091009 (2024).
25. M. Hino and T. Oda, *Hamon* **31**, 36 (2021).
26. T. Higuchi for the TUCAN Collaboration, *EPJ Web Conf.* **262**, 01015 (2022).
27. C. Abel *et al.*, *Phys. Rev. A* **99**, 042112 (2019).
28. W. Klassen *et al.*, *Eur. Phys. J. C* **84**, 1181 (2024).
29. M. McCrea for the TUCAN Collaboration, *Proceedings of Science*, PoS(PANIC2021), **459** (2022).

30. L.R. Hunter *et al.*, Proceedings of the Meeting on CPT and Lorentz Symmetry, World Scientific, Singapore, 1999.
31. C.J. Berglund *et al.*, *Phys. Rev. Lett.* **75**, 1879 (1995).
32. V.A. Kostelecký and C.D. Lane, *Phys. Rev. D* **60**, 116010 (1999).
33. S.K. Peck *et al.*, *Phys. Rev. A* **86**, 012109 (2012).
34. I. Altarev *et al.*, *Phys. Rev. Lett.* **103**, 081602 (2009).
35. I. Altarev *et al.*, *Physica B* **406**, 2365 (2011).
36. S.K. Lamoreaux *et al.*, *Phys. Rev. Lett.* **57**, 3125 (1986).
37. Y.V. Stadnik and V.V. Flambaum, *Eur. Phys. J. C* **75**, 110 (2015).
38. T.E. Chupp *et al.*, *Phys. Rev. Lett.* **63**, 1541 (1989).
39. *Data Tables for Lorentz and CPT Violation,* V.A. Kostelecký and N. Russell, 2025 edition, arXiv:0801.0287v18.
40. V.A. Kostelecký and A.J. Vargas, *Phys. Rev. D* **98**, 036003 (2018).

Chapter 21

New Directions in Gravity Searches for Spacetime-Symmetry Breaking

Q.G. Bailey

Department of Physics and Astronomy, Embry-Riddle Aeronautical University, Prescott, AZ 86301, USA

In this talk, we review recent results in testing spacetime symmetries in gravitational physics. Topics discussed include new signals for hypothetical Lorentz and diffeomorphism symmetry violations in short-range gravity tests. We review results for multipole expansions that predict extra polarizations for gravitational waves. Both explicit and spontaneous spacetime-symmetry-breaking origins are considered. We also discuss recent numerical results for black hole solutions in a vector field model of spontaneous symmetry breaking.

1. Introduction

In recent decades, a plethora of observational and experimental results have been obtained constraining spacetime-symmetry breaking in gravitational physics.[1–4] Tests include short-range gravity,[5] gravimetry,[6] lunar laser ranging,[7] pulsars,[8] and gravitational waves (GW).[9] While no statistically significant signal for spacetime-symmetry breaking has been found, motivation for these searches comes from the possibility of obtaining a hint of new physics beyond General Relativity (GR) and the Standard Model of particle physics. Indeed, mechanisms exist that could produce broken local Lorentz symmetry, diffeomorphism symmetry, and CPT symmetry.[10]

Many searches for spacetime-symmetry breaking use an effective field theory (EFT) framework with generic symmetry-breaking coordinate scalars added to the action of GR and the Standard Model.[11,12]

These terms have tensor-indexed coefficients controlling the degree of symmetry breaking. The premise is that spacetime-symmetry breaking is described by one or more background tensor fields coupled to known matter and fields. In other cases, specific "toy models," that can generate some of these background coefficients, are studied for hints on how symmetry-breaking mechanisms might work, including spontaneous symmetry breaking.

In this presentation, we briefly describe several recent areas for exploration in gravitational physics tests of spacetime symmetries with the author's direct involvement. The spacetime metric signature is taken as $-+++$, tensor indices are Greek letters μ, ν, ..., and other conventions can be found in the works discussed below.

2. Short-Range Gravity Tests

Despite many precision experiments, the nature of gravity is unknown on length scales less than micrometers. Proposals for "fifth" forces from strong gravity interactions abound. At submicrometer scales, these forces could be much stronger than the Newtonian gravitational force, and yet be consistent with current experimental limits.[13]

Lorentz-symmetry-breaking signals for short-range gravity tests were found using the EFT framework in the gravity sector. In Refs. 14 and 15, an approximation to first order in the coefficients was used to find the modified Newtonian force. For example, the modified Newtonian potential from a point mass m at the origin can be written in terms of spherical coefficients for symmetry breaking ($k_{jm}^{N(d)\text{lab}}$) as

$$U = \frac{G_N m}{r} + \sum_{djm} \frac{G_N m}{r^{d-3}} Y_{jm}(\theta,\phi) k_{jm}^{N(d)\text{lab}}, \tag{1}$$

where the angular dependence θ, ϕ is in local laboratory coordinates and d labels the mass dimension of the terms in the action.[16]

The lab-frame coefficients pick up time dependence when related to putative Sun-centered frame coordinates,[1] thereby serving as a signature for experiments. The result in Eq. (1) has already been used for analysis in experiments.[5] Recent results place limits on 14 $k_{jm}^{N(6)}$ coefficients and 22 $k_{jm}^{N(8)}$ coefficients at the $10^{-9}\,\text{m}^2$ and $10^{-12}\,\text{m}^4$ levels, respectively.

The first-order approximation makes searches in some short-range tests challenging, as some tests are designed to probe very small length scales at the cost of sensitivity to forces of Newtonian strength. In Ref. 17, the authors found exact solutions for a subset of coefficients in the EFT framework. Some of these coefficients are inaccessible in other tests and, furthermore, they allow for forces larger than the Newtonian force between masses on small length scales.

In the presence of certain forms of spacetime-symmetry breaking, the Newtonian potential between two masses depends on two length scales, and the relative amplitudes are controlled by the relative sizes of the coefficients. The result for the Newtonian potential Green function for a point mass at $\vec{r}'$ is

$$\mathcal{G}_1 = \frac{1}{2\pi R}\left[1 + \frac{1}{2}a_1 e^{\pm i w_1 R} - \frac{1}{2}a_2 e^{\pm i w_2 R}\right], \tag{2}$$

where $R = |\vec{r} - \vec{r}'|$. In this expression, the amplitudes of the terms depend on coefficient ratios, e.g.,

$$a_1 = \left(\frac{\pm(1+2\chi)}{\sqrt{1+4\chi}} - 1\right), \tag{3}$$

where $\chi = (k_2 + k_3)/k_1$ (the k_n's are short for rotational scalar coefficients like $k_1 = (k^{(6)})^{0i0i0j0j}/10$, etc.). The "wave numbers" w_1, w_2 depend also on the k_n's, leading to damping and oscillatory behavior.[17]

The new result is isotropic and so does not introduce sidereal dependence, but it introduces two length scales and possibly large-amplitude behavior (near $\chi = -1/4$) that could be measured in sensitive tests.[18] It remains an open problem for experimental analysis.

3. Gravitational Wave Generation and Polarizations

In one limit of the EFT approach, the terms in the Lagrange density for the gravity sector are constructed with the weak-field limit of gravity in mind.[16,19] Thus, with $g_{\mu\nu} = \eta_{\mu\nu} + h_{\mu\nu}$, the quadratic Lagrangian terms are

$$\mathcal{L} = -\frac{1}{4\kappa}\left(h^{\mu\nu}G_{\mu\nu} - \overline{s}^{\mu\kappa}h^{\nu\lambda}\mathcal{G}_{\mu\nu\kappa\lambda} + \tfrac{1}{4}h_{\mu\nu}q^{(5)\mu\rho\alpha\nu\beta\sigma\gamma}\partial_\beta R_{\rho\alpha\sigma\gamma} + \ldots\right), \tag{4}$$

where $\mathcal{G}$ is the double dual of the Riemann tensor and $\kappa = 8\pi G_N$.[14] The ellipsis denotes higher terms in the series organized by mass dimension d of the Lagrangian terms. The leading coefficients for Lorentz violation are the dimensionless $\overline{s}_{\mu\nu}$ coefficients and the coefficients $q^{(5)\mu\rho\alpha\nu\beta\sigma\gamma}$, the latter with mass dimension M^{-1}. These coefficients have been studied in solar-system tests, gravitational-wave tests, and others.

In a recent work, the author and collaborators studied the multipole expansion of the spacetime metric and curvature far away from a radiating source, focusing on the $\overline{s}_{\mu\nu}$ term in Eq. (4).[20] The basic equation solved takes the form

$$\hat{K}^{\mu\nu\alpha\beta} h_{\alpha\beta} = \kappa \tau^{\mu\nu}, \tag{5}$$

where $\hat{K}$ is a second-derivative operator with Riemann symmetries built from $\eta_{\mu\nu}$ and $\overline{s}_{\mu\nu}$ that satisfies $\partial_\mu \hat{K}^{\mu\nu\alpha\beta} = 0$. Also in this expression, $\tau^{\mu\nu}$ is the stress-energy pseudo-tensor.

Solutions to this equation have been obtained in a particular gauge: $(\eta^{\mu\nu} + \overline{s}^{\mu\nu})\partial_\mu h_{\nu\lambda} = \frac{1}{2}\partial_\lambda(\eta^{\mu\nu} + \overline{s}^{\mu\nu})h_{\mu\nu}$. At leading order in $\overline{s}_{\mu\nu}$, the result is

$$\begin{aligned} h_{\mu\nu} = \frac{\kappa}{2\pi\sqrt{-\tilde{g}}} \int d^3r' \frac{1}{\tilde{R}} \Big(&\tau_{\mu\nu} - 2\tau^\alpha{}_{(\mu}\overline{s}_{\nu)\alpha} - \frac{1}{2}\eta_{\mu\nu}(\tau^\alpha{}_\alpha - \overline{s}_{\alpha\beta}\tau^{\alpha\beta}) \\ &+ \frac{1}{2}\overline{s}_{\mu\nu}\tau^\alpha{}_\alpha \Big)(\tilde{t}_R, \vec{r}'), \end{aligned} \tag{6}$$

where $\tilde{t}_R$ and $\tilde{R}$ are modified retarded time and distance according to $\tilde{t}_R = t - (\tilde{R} - \overline{s}_{0i}R^i)/(1 + \overline{s}_{00})$, $\tilde{R} = \sqrt{(\delta_{ij} - \overline{s}_{ij})R^iR^j + O(\overline{s}^2)}$, and $R^i = r^i - r'^i$. The modified metric is $\tilde{g}^{\mu\nu} = \eta^{\mu\nu} + \overline{s}^{\mu\nu}$.[20]

In terms of physically measurable quantities, we focus on the spacetime curvature components relevant for GW measurements; namely, the six components R_{0i0j}.[2] These curvature components are calculated in the wave zone and to lowest order in a weak-field, slow-motion expansion appropriate for gravitational waves.[21] We obtain answers to the same order as the usual quadrupole results of GR (of order $(v/c)^4$ in h_{ij} in GR). The results are dependent on the inertia tensor $I_{jk} = \int d^3r\, \tau^{00} r^i r^j$. In GR, for a wave traveling in the $z = x^3$ direction, there are two independent polarizations corresponding to $R_{0202} - R_{0101}$ ("plus") and R_{0102} ("cross"), where x^1 and x^2 are the coordinates for the two transverse directions.

In theories beyond GR, up to four more polarizations can exist. For the beyond-GR polarizations, we find a "breathing" mode, and two "vector" modes given by

$$\begin{aligned} R_{0101} + R_{0202} &= \frac{G}{\tilde{r}} \left[(\overline{s}_{tr})_{\perp ij} + \frac{1}{2}(\overline{s}_{tr})_{nn}(\delta_{ij} - n_i n_j) \right] (\overset{(4)}{I})^{ij}, \\ R_{030i} &= \frac{G}{\tilde{r}} \big[\tfrac{1}{2}\big((\overline{s}_{tr})_{in} + \overline{s}_{0i}\big)(\delta_{jk} - n_j n_k)(\overset{(4)}{I})^{jk} \\ &\quad + \big((\overline{s}_{tr})_{nk_\perp} + \overline{s}_{0k_\perp}\big)(\overset{(4)}{I})^{ik} \big], \end{aligned} \tag{7}$$

where projections of quantities along $\hat{n} = \vec{r}/r$ are denoted with the index n and $\perp$ indicates a projection perpendicular to $\hat{n}$. The distance $\tilde{r}$ is obtained from $\tilde{R}$ with $\vec{r}' = 0$, $(\overline{s}_{tr})_{ij} = \overline{s}_{ij} - \delta_{ij}\overline{s}_{kk}/3$, and the (4) overset indicates the fourth $\tilde{t}_r$ derivative of the inertia tensor. The results above are evaluated at the modified retarded time $\tilde{t}_r$ (the $\vec{r}' = 0$ limit of $\tilde{t}_R$ above).

Additional polarizations can be sought in existing and future GW data. Some recent works have achieved results with Bayesian analysis favoring tensor over scalar modes and obtaining direct numerical bounds on the scalar-tensor ratio.[22] Other possibilities for extraction of polarization content include a null-stream method.[23] For a more complete list of references on this topic, see Refs. 24 and 25.

The analysis above was only for the $\overline{s}_{\mu\nu}$ coefficients, a subset of the more complete action in Eq. (4). A thorough program for the coefficients for any mass dimension has recently been countenanced.[26] Tests with data from the space interferometer mission LISA are also planned for the future.[27]

4. Results for Explicit Symmetry Breaking

The author and collaborators have also studied the explicit breaking limit of the $s_{\mu\nu}$ coefficients and other terms in the EFT action that are not in the linearized approximation in Refs. 24 and 28. Other authors have also studied this scenario.[29,30] In Ref. 24, the explicit breaking limit of the EFT considered is described by the Lagrange density

$$\begin{aligned}\mathcal{L} = \frac{\sqrt{-g}}{2\kappa}\Big[&R\left(1 - e_1 u + e_2 s^\alpha{}_\alpha + e_3 t^{\alpha\beta}{}_{\alpha\beta}\right)\\ &+ R^{\mu\nu}\left(e_4 s_{\mu\nu} + e_5 t_\mu{}^\alpha{}_{\nu\alpha}\right) + e_6 t_{\alpha\beta\mu\nu}R^{\alpha\beta\mu\nu}\Big] + \mathcal{L}_M,\end{aligned} \tag{8}$$

where the e_n's are dimensionless constants; they are introduced to distinguish the contributions from independent trace terms. The coefficients $s_{\mu\nu}$ and $t_{\alpha\beta\gamma\delta}$ are defined in Ref. 12. It is assumed that any matter ($\mathcal{L}_M$) is not coupled to these coefficients.

In the explicit breaking context, the traced Bianchi identities $\nabla_\mu G^{\mu\nu} = 0$ impose severe constraints on the field equations via $\nabla_\mu T^{\mu\nu} = 0$, since there are no dynamical equations for the coefficients.[31] It turns out that certain choices of the constants in Eq. (8) can circumvent the constraints stemming from the Bianchi identities. For example, if one chooses $e_4 - e_6 = 0$ then the constraint equation ($\nabla_\mu T^{\mu\nu} = 0$) can be satisfied if the Ricci scalar vanishes. Furthermore, one is then left with only a scalar combination

$\Phi = -e_1 u + e_2 s^\alpha{}_\alpha + e_3 t^{\alpha\beta}{}_{\alpha\beta}$. In fact, the field equations are then

$$-G^{\mu\nu}(1+\Phi) - (g^{\mu\nu}\nabla^2 - \nabla^\mu\nabla^\nu)\Phi + 8\pi G_N T^{\mu\nu} = 0. \tag{9}$$

The trace of this equation gives $3\nabla^\mu\nabla_\mu\Phi = 8\pi G_N T^\alpha{}_\alpha$, which is a wave equation for an extra scalar degree of freedom, arising here from explicit diffeomorphism breaking. These field equations are identical to a special limit of the Brans–Dicke model with $\omega = 0$,[2] with this limit ruled out by gravitational measurements.

If one calculates the effects of this scalar mode on gravitational waves, as in the previous section, an extra breathing mode is obtained. Specifically, in addition to the two usual plus and cross modes, the curvature obtains the added contribution

$$R_{0j0k} \supset \frac{G_N}{6r}(\delta_{jk} - n_j n_k)\partial_t^4(n_l n_m I^{lm} - I^{ll}). \tag{10}$$

The missing small-coefficient prefactor in this expression indicates no smooth limit with GR, hence a discontinuity, as in massive gravity.[32] There is also a general discussion on the absence of a smooth GR limit in Ref. 30.

When allowing for $e_4 \neq 0$, and an arbitrary background $s_{\mu\nu}$, GW solutions for Eq. (8) can be found that are not immediately ruled out by measurements. These results have been worked out in Ref. 24, and the results do show the persistence of an unsuppressed breathing mode, but the results are more complicated than the scalar results above, having dependence on various components of the coefficients $s_{\mu\nu}$.

5. Bumblebee Black Holes

An alternative approach to understanding spontaneous symmetry breaking is to study specific models.[33–37] Vector models of spontaneous symmetry breaking have been studied extensively in the last decades; for example, see Ref. 38. The Lagrange density for such models is

$$\mathcal{L} = \sqrt{-g}\left[\tfrac{1}{2\kappa}R - \frac{1}{4}B^{\mu\nu}B_{\mu\nu} - V + \ldots\right] + \mathcal{L}_M, \tag{11}$$

where the ellipsis includes nonminimal couplings to curvature, current couplings, etc.

Recently, a black-hole solution was found in Ref. 39 that arises from a nonminimal coupling to gravity ($\mathcal{L} \sim \xi B_\mu B_\nu R^{\mu\nu}$). This led to many follow-up papers, including ones with numerical solutions.[a] In nearly all these works, it is assumed that the vector B_μ is constrained to lie at the minimum of the symmetry-breaking potential $V(B^\mu B_\mu)$, thus $V' = 0$, where the prime is the derivative with respect to the argument.

The author and collaborators studied the case where $V' \neq 0$, but with the assumption of vanishing nonminimal couplings.[41] One motivation for this study was to explore gravity solutions when the potential V takes the special form of Kummer hypergeometric functions $M(n, 2, z)$ with $z = B_\mu B^\mu/\Lambda^2$; such potentials have been shown to yield renormalizable and stable quantum field theories in Refs. 37 and 42. Included in this work is a flat spacetime study of the modified classical electrodynamics of the model.[43] For brevity, we focus on the solutions for the spacetime metric and vector field in a spherically symmetric and static spacetime.

For the study of black holes we used Eddington–Finkelstein coordinates (v, r, θ, ϕ) where the metric line element takes the form $ds^2 = -N(r)\,dv^2 + 2M\,dv\,dr + r^2 d\Omega^2$. These coordinates are useful since one can define a horizon by $N(r_h) = 0$ for some radial coordinate value r_h. In these coordinates the bumblebee field is $B_\mu = (B_v, B_r, 0, 0)$. A version of the field equations in this metric, where the variables have been scaled to be dimensionless, is convenient for numerical solutions. These equations take the form

$$
\begin{aligned}
f'' + \frac{2}{x} f' &= 2\tilde{V}' \frac{M}{N} f + k \tilde{V}' \frac{M^2}{N^2} f^2 f' x, \\
M' &= \kappa \frac{f^2}{N^2} M^3 \tilde{V}' x, \\
N' &= \frac{(M^2 - N)}{x} - \frac{\kappa}{2} (f')^2 x - \kappa M^2 \tilde{V} x,
\end{aligned}
\tag{12}
$$

where $x = r/r_h \geq 0$, the scaled vector field is $f = B_v/\Lambda$, and $k = \kappa\Lambda^2$, with Λ being a suitable energy scale for the bumblebee field. The dimensionless potential terms are $\tilde{V} = (r_h^2/\Lambda^2)V$ and $\tilde{V}' = r_h^2 V'$. The other bumblebee equation for B_r turns out to provide a constraint relating B_r to B_v for the case $V' \neq 0$, which is already accounted for in Eq. (12).

First, we examined exact solutions to these equations that exist for the case $M' = 0$, revealing Schwarzschild–de Sitter and Reissner–Nordström spacetimes.[41] Second, we studied numerical solutions using a near-horizon analytical expansion for the functions N, M, and f. Since the numerical

[a]See, for example, Refs. 40, and a more complete list in Ref. 41.

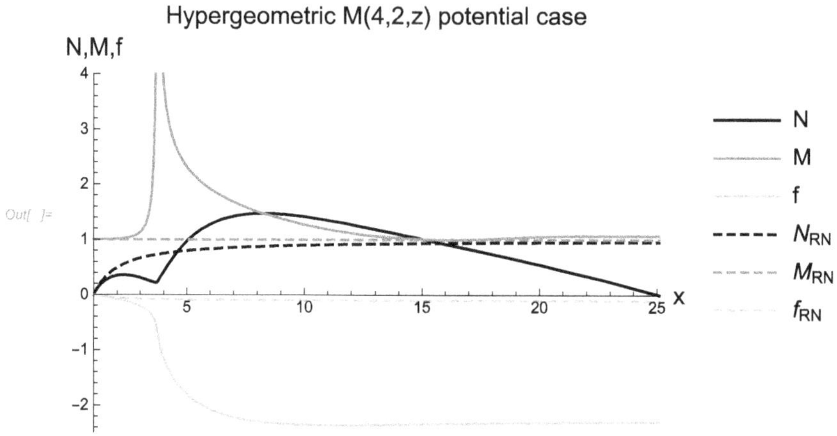

Fig. 1. The hypergeometric potential $\tilde{V} = \tilde{g}(M(4,2,z) + \frac{1}{2e^2})$ case. The initial horizon value used is $f'(1) = -1/9$, while $\tilde{g} = 1/8$.

code fails for $N \to 0$, we use a series expansion for the functions N, M, and f to seed numerical solutions that start slightly away from $x = 1$ (the horizon).

Highlights of our findings include peculiar behavior of the metric functions. We studied the quadratic potential, and two hypergeometric potentials. We display a sample in Fig. 1. Also in Ref. 41, we used a series expansion method for the asymptotic region $x \to \infty$ with the variable change $u = 1/x$. These results showed that for various potential choices, naked singularity solutions and solutions with peculiar horizon-like singularities occur. It remains an open problem to study these results in more detail.

Acknowledgments

For the work described in this proceedings contribution, Q.G. Bailey was supported by National Science Foundation grants 2207734 and 2308602.

References

1. V.A. Kostelecký and N. Russell, Data tables for Lorentz and CPT violation, *Rev. Mod. Phys.* **83**, 11 (2011); 2025 edition, arXiv:0801.0287v18.
2. C.M. Will, *Theory and Experiment in Gravitational Physics,* Cambridge University Press, Cambridge, England, 2018.
3. T. Mariz *et al.*, *Lorentz Symmetry Breaking—Classical and Quantum Aspects,* Springer, SpringerBriefs in Physics, DOI: 10.1007/978-3-031-20120-2, (2023).

4. A. Addazi *et al.*, Quantum gravity phenomenology at the dawn of the multi-messenger era—A review, *Prog. Part. Nucl. Phys.* **125**, 103948 (2022).
5. J.C. Long and V.A. Kostelecký, *Phys. Rev. D* **91**, 092003 (2015); C.G. Shao *et al.*, *Phys. Rev. D* **91**, 102007 (2015); *Phys. Rev. Lett.* **117**, 071102 (2016); *Phys. Rev. Lett.* **122**, 011102 (2019).
6. H. Müller *et al.*, *Phys. Rev. Lett.* **100**, 031101 (2008); K.-Y. Chung *et al.*, *Phys. Rev. D* **80**, 016002 (2009); V.A. Kostelecký and J.D. Tasson, *Phys. Rev. D* **83**, 016013 (2011); N.A. Flowers *et al.*, *Phys. Rev. Lett.* **119**, 201101 (2017); C.-G. Shao *et al.*, *Phys. Rev. D* **97**, 024019 (2018).
7. A. Bourgoin *et al.*, *Phys. Rev. Lett.* **117**, 24130 (2016); *Phys. Rev. Lett.* **119**, 201102 (2017); *Phys. Rev. D* **103**, 065055 (2021).
8. L. Shao, *Phys. Rev. Lett.* **112**, 111103 (2014); *Phys. Rev. D* **90**, 122009 (2014); L. Shao and Q.G. Bailey, *Phys. Rev. D* **98**, 084049 (2018); *Phys. Rev. D* **99**, 084017 (2019); N. Wex and M. Kramer, *Universe* **6**, 156 (2020).
9. B.P. Abbott *et al.*, *Astrophys. J.* **848**, L13 (2017); L. Shao, *Phys. Rev. D* **101**, 104019 (2020); L. Haegel *et al.*, *Phys. Rev. D* **107**, 064031 (2023).
10. V.A. Kostelecký and S. Samuel, *Phys. Rev. D* **39**, 683 (1989); R. Gambini and J. Pullin, *Phys. Rev. D* **59**, 124021 (1999).
11. V.A. Kostelecký and R. Potting, *Phys. Rev. D* **51**, 3923 (1995); D. Colladay and V.A. Kostelecký, *Phys. Rev. D* **55**, 6760 (1997); *Phys. Rev. D* **58**, 116002 (1998).
12. V.A. Kostelecký, *Phys. Rev. D* **69**, 105009 (2004).
13. J. Murata and S. Tanaka, *Class. Quant. Grav.* **32**, 033001 (2015).
14. Q.G. Bailey *et al.*, *Phys. Rev. D* **91**, 022006 (2015).
15. V.A. Kostelecký and M. Mewes, *Phys. Lett. B* **766**, 137 (2017).
16. V.A. Kostelecký and M. Mewes, *Phys. Lett. B* **757**, 510 (2016).
17. Q.G. Bailey *et al.*, *Class. Quant. Grav.* **40**, 045006 (2023).
18. R.S. Decca *et al.*, *Phys. Rev. Lett.* **94**, 240401 (2005).
19. Q.G. Bailey and V.A. Kostelecký, *Phys. Rev. D* **74**, 045001 (2006).
20. Q.G. Bailey *et al.*, *Ann. Phys.* **461**, 169582 (2024).
21. L. Blanchet, *Living Rev. Rel.* **17**, 2 (2014).
22. H. Takeda *et al.*, *Phys. Rev. D* **103**, 064037 (2021); *Phys. Rev. D* **105**, 084019 (2022).
23. D. Liang *et al.*, *Phys. Rev. D* **110**, 084040 (2024).
24. Q.G. Bailey *et al.*, *Phys. Rev. D* **110**, 084066 (2024).
25. N. Yunes *et al.*, *Living Rev. Rel.* **28**, 3 (2025).
26. N.A. Nilsson and C. Le Poncin-Lafitte, *Phys. Rev. D* **109**, 024035 (2024); S.A. Lafkih *et al.*, arXiv:2506.08859.
27. M. Colpi *et al.*, *LISA Definition Study Report,* arXiv:2402.07571.
28. K. O'Neal-Ault *et al.*, *Phys. Rev. D* **103**, 044010 (2021).
29. R. Bluhm, *Phys. Rev. D* **91**, 065034 (2015); Y. Bonder, *Phys. Rev. D* **91**, 125002 (2015); C.M. Reyes and M. Schreck, *Phys. Rev. D* **104**, 124042 (2021); Y. Bonder and C. Peterson, *Phys. Rev. D* **101**, 064056 (2020); C. Reyes *et al.*, *Phys. Rev. D* **111**, 124011 (2025).
30. V.A. Kostelecký and Z. Li, *Phys. Rev. D* **103**, 024059 (2021).
31. R. Bluhm *et al.*, *Phys. Rev. D* **100**, 084022 (2019).

32. K. Hinterbichler, *Rev. Mod. Phys.* **84**, 671 (2012).
33. V.A. Kostelecký and S. Samuel, *Phys. Rev. D* **39**, 683 (1989).
34. V.A. Kostelecký and R. Lehnert, *Phys. Rev. D* **63**, 065008 (2001).
35. T. Jacobson and D. Mattingly, *Phys. Rev. D* **64**, 024028 (2001).
36. V.A. Kostelecký and R. Potting, *Phys. Rev. D* **79**, 065018 (2009).
37. B. Altschul *et al.*, *Phys. Rev. D* **81**, 065028 (2010).
38. R. Bluhm and V.A. Kostelecký, *Phys. Rev. D* **71**, 065008 (2005); C. Eling and T. Jacobson, *Class. Quant. Grav.* **23**, 5643 (2006); S.M. Carroll *et al.*, *Phys. Rev. D* **79**, 065011 (2009); M. Seifert, *Phys. Rev. D* **79**, 124012 (2009); C. Hernaski, *Phys. Rev. D* **90**, 124036 (2014); A. Delhom *et al.*, JCAP **07**, 018 (2022).
39. R. Casana *et al.*, *Phys. Rev. D* **97**, 104001 (2018).
40. R.V. Maluf and J.C.S. Neves, *Phys. Rev. D* **103**, 044002 (2021); R. Xu *et al.*, *Phys. Rev. D* **107**, 024011 (2023); A.A.A. Filho *et al.*, *Phys. Rev. D* **108**, 085010 (2023); Z.-F. Mai *et al.*, *Phys. Rev. D* **109**, 084076 (2024).
41. Q.G. Bailey, R. Murray, and D. Walter-Cardona, *Phys. Rev. D* **112**, 024069 (2025).
42. B. Altschul and V.A. Kostelecký, *Phys. Lett. B* **628**, 106 (2005).
43. D. Walter-Cardona, these proceedings.

Chapter 22

Features of Spacetime-Symmetry Breaking and the Standard-Model Extension in Riemann–Cartan Geometry

R. Bluhm

Department of Physics and Astronomy, Colby College, Waterville, ME 04901, USA

For over two decades, the gravity sector of the Standard-Model Extension (SME) has served as a phenomenological framework for testing spacetime symmetry breaking in the presence of gravity. During this time, various theoretical features have been examined in greater detail and some refinements have been made. In particular, differences between spontaneous and explicit breaking of diffeomorphisms, local translations, and local Lorentz transformations in Riemann-Cartan geometry, as well as their corresponding consistency issues with geometric and mathematical identities, have been probed more deeply. This has led to a modified version of the SME being developed that is suitable for investigating explicit breaking in gravity theories, which can be used as well to search for new geometries that go beyond Riemann-Cartan. A selective overview of some of these features is presented here.

1. Introduction

The gravity sector of the Standard-Model Extension (SME) contains all observer-invariant terms that couple gravitational and matter fields with fixed background fields associated with spacetime symmetry breaking.[1] It is naturally defined using a vierbein formalism to accommodate fermion fields with spin, where the independent gravitational fields are the vierbein ${e_\mu}^a$ and spin connection ${\omega_\mu}^{ab}$. The spacetime can then have both curvature and torsion, and the geometry is Riemann–Cartan. The generic form of the action with an Einstein–Hilbert term can be written as

$$S = \frac{1}{2\kappa} \int d^4x \, e \, R({e_\mu}^a, {\omega_\mu}^{ab}) + \int d^4x \, e \, \mathcal{L}_{g,m,\bar{k}_X}({e_\mu}^a, {\omega_\mu}^{ab}, f^\psi, \bar{k}_X), \quad (1)$$

with $\kappa = 8\pi G$, and where f^ψ represents matter fields with indices represented by ψ, while $\bar{k}_X$ is a fixed background with indices represented by X. In each of these, the indices can be combinations of spacetime indices (denoted as μ, ν, ...) and local indices (a, b, ...).

The Einstein, torsion, and matter field equations of motion are obtained, respectively, by varying the action with respect to ${e_\mu}^a$, ${\omega_\mu}^{ab}$, and f^ψ, and are given as

$$G^{\mu\nu} = \kappa \bar{T}_e^{\mu\nu}, \quad (2)$$

$$\hat{T}^{\lambda\mu\nu} = -\kappa {\bar{S}_\omega}{}^{\lambda\mu\nu}, \quad (3)$$

$$\frac{\delta S_{g,m,\bar{k}_X}}{\delta f^\psi} = 0, \quad (4)$$

where

$$\hat{T}^{\lambda\mu\nu} = T^{\lambda\mu\nu} + {T^\alpha}_\alpha{}^\mu g^{\lambda\nu} - {T^\alpha}_\alpha{}^\nu g^{\lambda\mu} \quad (5)$$

is the trace-corrected torsion tensor, $\bar{T}_e^{\mu\nu}$ is the energy-momentum, and ${\bar{S}_\omega}{}^{\lambda\mu\nu}$ is the spin density. The bars over the two latter quantities denote that they depend on the fixed background $\bar{k}_X$.

As fixed backgrounds, the fields $\bar{k}_X$ do not transform under diffeomorphisms or particle Lorentz transformations, and as a result these symmetries are broken either explicitly or spontaneously. In the case of explicit breaking, the backgrounds $\bar{k}_X$ are understood as completely nondynamical objects that are put directly into the action, while in the case of spontaneous breaking the backgrounds $\bar{k}_X$ arise as vacuum values of fully dynamical fields K_X, e.g., $\bar{k}_X = \langle K_X \rangle$.

To maintain observer independence, the action S must be invariant under general coordinate transformations and observer local Lorentz transformations. These invariances give rise to mathematical Noether identities that hold off shell. In addition, geometric Bianchi identities involving the curvature and torsion must always hold. Notice that the backgrounds $\bar{k}_X$ are not varied in the equations of motion shown above; however, they do transform under general coordinate transformations and observer Lorentz transformations.

With the absence of variations of $\bar{k}_X$ in the equations of motion, conflicts can arise between the off-shell Noether or Bianchi identities and the on-shell equations of motion for the vierbein, spin connection, and matter fields. This is highly problematic for the case of explicit breaking, because the variations with respect to $\bar{k}_X$ need not vanish, so that

$$\frac{\delta S_{\bar{k}_X}}{\delta \bar{k}_X} \neq 0 \,. \tag{6}$$

In this case, theoretical inconsistency, referred to as no-go results, with Riemann–Cartan geometry can occur.[1] In contrast, with spontaneous breaking, the variations of the action with respect to the dynamical fields K_X for vacuum solutions $\langle K_X \rangle = \bar{k}_X$ do vanish,

$$\frac{\delta S_{\bar{k}_X}}{\delta \bar{k}_X} = 0 \,, \tag{7}$$

since these are vacuum solutions, where $e_\mu{}^a$ and $\omega_\mu{}^{ab}$ also equal their vacuum values $\langle e_\mu{}^a \rangle$ and $\langle \omega_\mu{}^{ab} \rangle$. Thus, the no-go results are evaded for the case of spontaneous breaking. It is for this reason that the backgrounds in the original SME gravity sector, referred to as SME coefficients, are considered as arising from spontaneous breaking, whereas explicit breaking that is inconsistent in Riemann–Cartan geometry can be viewed as an indication of beyond-Riemann–Cartan geometry, such as Finsler geometry.[1]

Many experimental tests of spontaneous Lorentz and diffeomorphism violation have been made using the SME,[2,3] and theoretical questions concerning, for example, the fate of the Nambu–Goldstone (NG) modes were considered early on.[4] Since then, many other theoretical features of both explicit and spontaneous breaking have been explored, which have implications for the gravity sector of the SME. A very selective overview of some of these is given in the following sections.

2. Explicit Breaking Revisited

Interest in theoretical models with explicit spacetime symmetry breaking, such as in Hořava gravity or massive gravity,[5] prompted a re-evaluation of the no-go results for explicit breaking.[6] In a nutshell, it was found that in some cases the no-go results can be evaded. This is because even though the variations in Eq. (6) do not hold as equations of motion for fixed nondynamical backgrounds, the left-hand side of Eq. (6) can nonetheless vanish for other reasons. One is that with explicit breaking, there are more degrees of freedom, due to the loss of local gauge symmetries. Thus, there are additional vierbein and/or spin connection modes compared to the case when the spacetime symmetries are not broken. Alternatively, the vanishing of the expression in Eq. (6) can hold if certain constraints are imposed on geometrical quantities, such as the curvature.[6–8]

In general, the counting of degrees of freedom shows that the no-go results can be evaded at least in principle at nonlinear order. However, in some cases, as with an ansatz solution for the metric, the extra modes do not appear. Similarly, working in perturbation theory can cause the extra modes to be suppressed at linear order, leading to inconsistency in low-energy effective theories. Likewise, in theories where the no-go results impose constraints on the curvature, these conditions can often be so severe as to make a theory nonviable. Thus, questions of consistency remain paramount in models with explicit breaking.

With these considerations in mind, a new version of the SME for the case of explicit breaking has been developed.[9] Its action generalizes the original SME action to include all observer-independent terms coupling gravitational and matter fields with fixed *nondynamical* backgrounds that explicitly break spacetime symmetries. General arguments are made showing that at the level of perturbation theory the new terms in the action lead to inconsistency in Riemann–Cartan geometry. Thus, the interpretation with the new version of the SME is that it can be used as a phenomenological framework in gravity tests looking for explicit breaking, but any detection of a signal would likely indicate a new geometry that goes beyond Riemann–Cartan.

An important caveat in using the new SME framework with explicit breaking is that couplings to different components of a background field must be treated as independent interactions. For example, for the case of a nondynamical background vector, terms in the action coupling to spacetime components $\bar{k}_\mu$ and $\bar{k}^\mu$ are physically distinct and must be investigated separately. This is because if both $\bar{k}_\mu$ and $\bar{k}^\mu$ are nondynamical they cannot be related by the physical metric $g^{\mu\nu}$. Similarly, local components $\bar{k}_a$ cannot be treated as being physically related to spacetime components $\bar{k}_\mu$, because the dynamical vierbein $e_\mu{}^a$ cannot connect two nondynamical quantities. However, since the local metric η_{ab} is itself fixed, couplings to local components $\bar{k}^a$ and $\bar{k}_a$ can be treated as physically equivalent. Notice that this caveat can be ignored for the case of spontaneous breaking in the context of the original SME because there the backgrounds are vacuum values, which are physically linked by the vacuum values for the dynamical vierbein, e.g., $\bar{k}_\mu = \langle e_\mu{}^a \rangle \, \bar{k}_a$.

As a result of the differences between nondynamical components, e.g., $\bar{k}_\mu$, $\bar{k}^\mu$, and $\bar{k}_a$, the phenomenological framework resulting from the SME with explicit breaking has many more terms and coefficients that need to be tested compared to the SME with dynamical backgrounds resulting from spontaneous breaking.[9] Examples in cosmology with explicit breaking have also recently been considered.[10,11]

3. Local Translations

Gravity theories in Riemann–Cartan geometry have three relevant local symmetries: diffeomorphisms, local translations, and local Lorentz transformations. However, only two of these are independent, since suitable combinations of diffeomorphisms and local Lorentz transformations can reproduce local translations, and vice versa.[12–14] Poincaré gauge theories of gravity work with translations and Lorentz transformations as the fundamental symmetries, while the gravity sector of the SME uses diffeomorphisms and Lorentz transformations as the independent pair of local symmetries. A natural question that then arises is whether the breakings of these local symmetries are related as well.

With spontaneous breaking of diffeomorphisms and local Lorentz invariance, it must be remembered that the symmetries are not actually broken, but rather they become hidden with respect to vacuum solutions that do not obey the symmetries. In this case, when a tensor has a vacuum value with spacetime components that break diffeomorphisms, the vacuum vierbein can be used to obtain the local components for the tensor, which spontaneously break local Lorentz transformations.[4] This is true as well for nonconstant scalars, which have derivatives along spontaneously chosen directions in both the spacetime and local frames.

Since local translations can be formed as combinations of diffeomorphisms and local Lorentz transformations, any tensor vacuum value that spontaneously breaks these symmetries will also spontaneously break local translations. Hence, any tensor background that arises from spontaneous breaking will break all three local spacetime symmetries. Note in particular that the vacuum value for the vierbein, $\langle e_\mu{}^a \rangle$, has both a spacetime and a local component, so it by itself spontaneously breaks all three local symmetries.

In contrast, with explicit breaking, a nondynamical tensor is included directly in the action, which can break spacetime symmetries. In this case, the three local spacetime symmetries need not all break if one of them breaks.[15,16] This is largely due to the fact that the physical vierbein cannot relate spacetime and local components of a fixed background. For example, for a fixed vector, theories with couplings to components $\bar{k}_\mu$, $\bar{k}^\mu$, and $\bar{k}_a$ must be treated as independent theories, as described previously, and thus can cause explicit breaking of different symmetries. For example, constant components $\bar{k}_\mu$ explicitly break diffeomorphisms and local translations, but not local Lorentz transformations, while constant components $\bar{k}_a$ explicitly break local translations and local Lorentz transformations, but not diffeomorphisms.

4. Nondynamical Scalar Backgrounds

Explicit breaking with fixed nondynamical scalar background fields gives rise to some noteworthy features.[6,7] For example, consider a gravity theory with a set of fixed nondynamical scalars, $\bar{\phi}^A(x)$, with $A = 1, 2, \ldots, N$. The presence of such scalars in the action explicitly breaks diffeomorphism invariance. However, to be physically viable the theory must still be observer independent, which means the action must be unchanged under general coordinate transformations, $x^\mu \to x^\mu + \xi^\mu$, where ξ^μ are arbitrary infinitesimal spacetime vectors. Mathematical identities following from this are at odds with the explicit diffeomorphism breaking and Bianchi identities, leading to no-go results, unless the following condition holds:[6]

$$\left[-D_\mu \frac{\partial \mathcal{L}}{\partial(\partial_\mu \bar{\phi}^A)} + \frac{\partial \mathcal{L}}{\partial \bar{\phi}^A}\right] \partial_\nu \bar{\phi}^A = 0 \,. \tag{8}$$

Assuming that the scalars are not constant, so that $\partial_\nu \bar{\phi}^A \neq 0$, and assuming their derivatives are linearly independent as well, it follows that a stricter set of conditions must hold:

$$-D_\mu \frac{\partial \mathcal{L}}{\partial(\partial_\mu \bar{\phi}^A)} + \frac{\partial \mathcal{L}}{\partial \bar{\phi}^A} = 0 \,. \tag{9}$$

Note that these have the form of the Euler–Lagrange equations that would hold dynamically if the scalars $\bar{\phi}^A$ were instead dynamical. Evidently, to evade the no-go results, the Euler–Lagrange equations for the scalars must still hold despite the fact that the scalars are not dynamical.

Equation (9) cannot follow from dynamical field variations of $\bar{\phi}^A$, since the scalars are fixed and nondynamical. Instead, Eq. (9) must hold for one of two reasons.

The first is that the covariant derivatives in Eq. (9) can involve additional degrees of freedom in the metric because of the loss of diffeomorphism invariance. With breaking associated with four local vectors ξ^μ, this would allow up to four additional modes, which matches the number of equations in Eq. (9) if $N = 4$. This result helps explain why the Stückelberg trick works in theories with explicit diffeomorphism breaking.[17] In the Stückelberg approach, a fixed background tensor is replaced by derivatives of four fixed Stückelberg scalar fields, $\bar{\phi}^A$, with $A = 0, 1, 2, 3$. The scalars are then promoted to dynamical fields with equations of motion given by their Euler–Lagrange equations, which happen to match the conditions for evading the no-go results in Eq. (9). Thus, as long as the Stückelberg scalars obey their equations of motion, then the no-go results are evaded. However, this is not guaranteed. A theory with Stückelberg fields can still be inconsistent if solutions to Eq. (9) do not exist.

A second way for the conditions in Eq. (9) to hold is if a constraint is imposed on the spacetime geometry. An example of this is provided by Chern–Simons gravity in four dimensions, which includes a nondynamical scalar in the action.[7] The condition for evading the no-go result in this case, which is equivalent to the Euler–Lagrange equation for the scalar, is that the gravitational Pontryagin density must vanish, $\star RR = R^{\alpha}{}_{\beta\mu\nu} R^{\beta}{}_{\alpha}{}^{\mu\nu} = 0$.

A nondynamical scalar can also be used to construct theories with a hybrid form of spacetime symmetry breaking involving both explicit and spontaneous breaking.[18,19] This occurs when a fixed nondynamical scalar is included in a potential that causes spontaneous breaking. For example, in bumblebee models with a vector field B_μ, spontaneous breaking is induced when the potential has the form $V(B^\mu B_\mu + b^2)$ with b equal to a constant. A vacuum solution $\langle B_\mu \rangle = b_\mu$ occurs, with a timelike vacuum value that can be written as $b_\mu = (b, 0, 0, 0)$ in an appropriate frame. In this case, NG modes are excited that stay in the minimum of the potential with $V' = 0$, while a massive mode occurs when $V' \neq 0$. By replacing the constant b by a nondynamical scalar in the potential V, for example using a time-dependent scalar $b(t)$ in $V(B^\mu B_\mu + b(t)^2)$, explicit diffeomorphism breaking occurs. However, the potential continues to induce spontaneous breaking but in this case with a time-dependent vacuum solution, $b_\mu(t) = (b(t), 0, 0, 0)$. The constraint that must hold to evade the no-go results is $V' = 0$, which also follows from the Euler–Lagrange equation for $b(t)$. This constraint eliminates the massive mode, leaving only NG modes as excitations in B_μ.

5. Spin Connection and Torsion Backgrounds

The roles of the spin connection and torsion have also been explored in greater detail in gravity theories with spacetime symmetry breaking.[15,16] Examples are given here for both explicit breaking as well as spontaneous breaking.

In models with explicit breaking, it has been shown that torsion can occur even when the spin density due to the matter fields is zero. This is different from the case of Einstein–Cartan theory in the absence of spacetime symmetry breaking, where the torsion $T^{\lambda\mu\nu}$ is directly proportional to the spin density $S_\omega{}^{\lambda\mu\nu}$. Thus, if the spin density vanishes then so does the torsion. However, with nondynamical background fields that cause explicit breaking, the torsion equation is modified, as shown in Eq. (3). Here, the bars over the torsion and spin density indicate interactions of the gravitational and matter fields with a nondynamical background. In models with pure-gravity couplings to the background, the torsion can be nonzero even in the absence of matter.[15]

In theories with spontaneous breaking of spacetime symmetries, it is common to assume that while the vierbein has a nonzero vacuum value, $\langle e_\mu{}^a \rangle \neq 0$, the spin connection has a vanishing vacuum value, $\langle \omega_\mu{}^{ab} \rangle = 0$. This is a simplifying assumption, since the torsion is related to the spin connection by

$$T_{\lambda\mu\nu} = e_\lambda{}^a \left[(\partial_\mu e_{\nu a} + \omega_{\mu ab}\, e_\nu{}^b) - (\mu \leftrightarrow \nu) \right] . \tag{10}$$

Thus, when the vierbein has a constant Minkowski vacuum value, e.g., $\langle e_\mu{}^a \rangle = \delta_\mu^a$, the torsion has a vanishing vacuum value whenever the spin connection also has a vanishing vacuum value. However, this assumption need not always hold. Models with $\langle \omega_\mu{}^{ab} \rangle \neq 0$ can also be considered, where in this case the torsion and spin density can both have nonzero vacuum values.

For example, consider a bumblebee model with torsion, where the field strength is given by

$$B_{\mu\nu} = D_\mu B_\nu - D_\nu B_\mu = \partial_\mu B_\nu - \partial_\nu B_\mu - T^\lambda{}_{\mu\nu} B_\lambda \, . \tag{11}$$

In a vacuum solution with a constant value of $\langle B_\mu \rangle = b_\mu$, the vacuum field strength becomes

$$\langle B_{\mu\nu} \rangle = -\langle T^\lambda{}_{\mu\nu} \rangle \, \langle e_\lambda{}^a \rangle \, \bar{b}_a \, , \tag{12}$$

the torsion vacuum value is

$$\langle T_{\lambda\mu\nu} \rangle = \langle \omega_{\mu\lambda\nu} \rangle - \langle \omega_{\nu\lambda\mu} \rangle \,, \tag{13}$$

and therefore $\langle \bar{S}_{\omega}{}^{\lambda\mu\nu} \rangle \neq 0$. Thus, if $\langle \omega_{\mu}{}^{ab} \rangle \neq 0$, nonzero vacuum values can occur for all of the quantities $\langle B_{\mu\nu} \rangle$, $\langle T_{\lambda\mu\nu} \rangle$, and $\langle \bar{S}_{\omega}{}^{\lambda\mu\nu} \rangle$ even in the absence of excitations in B_{μ}.

Vacuum values for the spin connection can also affect how the NG modes appear with spontaneous spacetime breaking. The NG modes originate as virtual broken symmetry transformations that stay in the potential minimum, which can consist of broken diffeomorphisms, local translations, or local Lorentz transformations. However, with a bumblebee potential only three NG excitations arise that stay in the minimum with $V' = 0$. The remaining degrees of freedom aside from the gravitational excitations and a massive mode are gauge fixed, and how this is carried out depends on whether $\langle \omega_{\mu}{}^{ab} \rangle$ vanishes or not. This is because b_{μ} by itself breaks diffeomorphisms but not local translations when $\langle \omega_{\mu}{}^{ab} \rangle = 0$; whereas with $\langle \omega_{\mu}{}^{ab} \rangle \neq 0$, b_{μ} by itself breaks both diffeomorphisms and local translations. Thus, with $\langle \omega_{\mu}{}^{ab} \rangle = 0$, and after gauge fixing, the NG modes are found to arise from b_{μ} as virtual local Lorentz transformations. In contrast, when $\langle \omega_{\mu}{}^{ab} \rangle \neq 0$ and after gauge fixing, the NG modes in this case appear as a locked combination of both local translations and local Lorentz transformations.[16]

References

1. V.A. Kostelecký, *Phys. Rev. D* **69**, 105009 (2004).
2. Q.G. Bailey and V.A. Kostelecký, *Phys. Rev. D* **74**, 045001 (2006); V.A. Kostelecký and J.D. Tasson, *Phys. Rev. D* **83**, 016013 (2011).
3. V.A. Kostelecký and N. Russell, *Rev. Mod. Phys.* **83**, 11 (2011).
4. R. Bluhm and V.A. Kostelecký, *Phys. Rev. D* **71**, 065008 (2005); R. Bluhm, S.-H. Fung, and V.A. Kostelecký, *Phys. Rev. D* **77**, 065020 (2008).
5. S. Mukohyama, *Class. Quantum Grav.* **27**, 223101 (2010); K. Hinterbichler, *Rev. Mod. Phys.* **84**, 671 (2012).
6. R. Bluhm, *Phys. Rev. D* **91**, 065034 (2015); R. Bluhm and A. Šehić, *Phys. Rev. D* **94**, 104034 (2016); R. Bluhm, H. Bossi, and Y. Wen, *Phys. Rev. D* **100**, 084022 (2019).
7. R. Jackiw, in *CPT and Lorentz Symmetry IV*, V.A. Kostelecký (ed.), World Scientific, Singapore, 2008.
8. K. O'Neal-Ault, Q.G. Bailey, and N.A. Nilsson, *Phys. Rev. D* **103**, 044010 (2021).
9. V.A. Kostelecký and Z. Li, *Phys. Rev. D* **103**, 024059 (2021); *Phys. Rev. D* **104**, 044054 (2021).

10. C.M. Reyes, M. Schreck, and A. Soto, *Phys. Rev. D* **106**, 023524 (2022).
11. N.A. Nilsson, *Phys. Rev. D* **106**, 104036 (2022).
12. F.W. Hehl, P. von der Heyde, G.D. Kerlick, and J.M. Nester, *Rev. Mod. Phys.* **48**, 393 (1976).
13. M. Blagojević, *Gravitation and Gauge Symmetries*, CRC Press, Boca Raton, FL, 2002.
14. M. Blagojević and F.W. Hehl, *Gauge Theories of Gravitation*, World Scientific, Singapore, 2013.
15. Y. Bonder and C. Corral, *Symmetry* **10**, 433 (2018).
16. R. Bluhm and Y. Zhi, *Symmetry* **16**, 25 (2024).
17. N. Arkani-Hamed, H. Georgi, and M.D. Schwartz, *Ann. Phys.* **305**, 96 (2003).
18. R. Bluhm, *Phys. Rev. D* **92**, 085015 (2015).
19. C.M. Reyes, C. Riquelme, M. Schreck, and A. Soto, *Phys. Rev. D* **111**, 124011 (2025).

Chapter 23

Status and Results of the Analysis for the UCNτ Neutron Lifetime Experiment

L.S. Blokland

Physics Department, Indiana University, Bloomington, IN 47405, USA

(On behalf of the UCNτ Collaboration)

This proceedings contribution outlines the methods and results of the analysis of the UCNτ experiment for data taken from 2020 to 2022. The UCNτ experiment is a neutron lifetime experiment at the Los Alamos Neutron Science Center, which has provided one of the most precise values for the neutron lifetime. Neutron beta decay is the simplest type of beta decay, and the neutron lifetime provides a useful way to probe the Standard Model, particularly by helping to determine V_{ud}, which can also be used to test the unitarity of the Cabibbo–Kobayashi–Maskawa matrix. For the UCNτ experiment, ultracold neutrons are stored in a magneto-gravitational bottle and later counted in situ by a ZnS alpha detector coated with enriched ^{10}B in order to extract the neutron lifetime.

1. Introduction

Neutron beta decay involves an up quark transitioning to a down quark through the emission of a virtual W boson. Measuring the neutron lifetime can be used as a test of the Standard Model because the lifetime can be used to calculate V_{ud} of the Cabibbo–Kobayashi–Maskawa (CKM) quark-mixing matrix using the relationship:[1]

$$\tau_n^{-1} = \frac{m_e^5}{2\pi^3} G_F^2 |V_{ud}|^2 (1 + 3g_A^2)(1 + RC) f. \tag{1}$$

Here, G_F is the Fermi coupling constant, g_A is the weak axial-current coupling, the RC term contains the radiative corrections, and f is a decay phase-space factor.

There are two main types of neutron beta decay experiments, which are called the "beam" method and the "bottle" method. In the beam method, decay protons from a beam of cold neutrons are trapped and counted, allowing for a measurement of the neutron lifetime.[2] In bottle-method experiments, such as UCNτ, ultracold neutrons are held in a trap for some holding period, after which the remaining neutrons in the trap are counted and the neutron lifetime is extracted.[3]

Experiments using these two methods have produced measurements of the neutron lifetime that do not agree with each other, with a 4σ difference between the lifetime values.[3] This disagreement could indicate either systematic effects in one or both methods that are unknown or not properly characterized, or potentially an unknown decay channel of the neutron.

2. Experimental Configuration

Spallation neutrons are produced at the Los Alamos Neutron Science Center (LANSCE) by a proton beam hitting a tungsten target. These neutrons are then moderated to cold temperatures using polyethylene and brought to ultracold temperatures using a solid deuterium crystal.[4] The UCN are then sent through guide tubes to the polarizing magnet, which allows neutrons in the "high-field-seeking" spin state, in which the neutrons have magnetic moments parallel to the magnetic field, to continue on, while neutrons in the "low-field-seeking" state, in which the neutrons have magnetic moments antiparallel to the magnetic field, are repelled. The polarized UCN then continue on to the "roundhouse," which is a cylindrical buffer volume made out of stainless steel and coated with NiP meant to smooth out effects due to fluctuations in the beam. After leaving the roundhouse, UCN are then flipped to the low-field-seeking state using a fast-passage adiabatic spin flipper. This is done so that the UCN can be trapped using magnetic fields.

After the spin flipper, UCN are then loaded into the trap through a trapdoor located at the bottom of the trap. The trap confines UCN from below using a Halbach array made up of neodymium magnets, while gravity keeps the UCN from escaping out the top of the trap. The trap is also surrounded by electromagnets to ensure the UCN do not encounter a region with zero magnetic field, which would cause them to lose their polarization. Once the UCN are loaded into the trap, the bottom of the trap is sealed and any higher-energy neutrons are removed from the trap using the spectral "cleaner." The UCN are then held in the trap for some holding period that ranges between 20 and 1550 seconds before the "dagger" detector is lowered to count the neutrons in situ. The dagger is a ZnS scintillator coated with ^{10}B connected to photomultiplier tubes (PMTs) using wavelength-shifting fibers. Three different configurations of the dagger detector were used for this set of data. For 2020 and the first part of 2021, both sides of the dagger were uncovered and two PMTs recorded the scintillation counts. For the second part of 2021, one side of the dagger was covered with aluminum to reduce the counting rate and reduce rate-dependent effects. For 2022, the dagger was split into four sections divided by aluminum strips, with each section having its own pair of PMTs. This was done to further reduce rate-dependent effects. After the remaining neutrons are counted, the dagger continues to take data in the trap for a period of time in order to obtain a background measurement.

3. Data Analysis

The most recent set of data taken from 2020 to 2022 has been analyzed, and the results published in Ref. 3. The data were analyzed by at least three separate teams for each year of data, with four total teams analyzing different combinations of years. The analysis was done separately for each year, with 2021 split further into two sets due to the change in the dagger detector during this year. Each year was also split into smaller groups based on running conditions for normalization purposes. The data also had to be screened so that runs done in poor conditions were removed. This was done by looking for abnormal conditions noted in the logbook, as well as by looking for runs with yields far off from the average of nearby runs with the same holding time.

The analysis consists of first reconstructing the neutron counts by looking for coincidences of photoelectrons in PMT pairs from the dagger detector. The UCN counts during the unload period where the dagger detector has lowered into the trap are then found for each run, along with a background measurement. The background-subtracted counts are then normalized using information from a monitor detector placed in the roundhouse, along with the neutron counts in the dagger detector during the filling of the trap.

In addition to the normalization, several systematic corrections had to be made to the lifetime. These corrections include statistical bias correction, pressure correction, and phase-space evolution correction. Additionally, a correction for higher-energy neutrons that are not removed from the trap during the cleaning period and a correction for neutrons that become vibrationally heated have to be made, as these effects can lead to neutrons escaping the trap during the holding period, thus shifting the lifetime. For the 2022 data, which used a different dagger from the 2020 and 2021 data, there is also an additional correction that stems from varying efficiencies of the dagger segments coupled with a phase-space evolution.

The normalized yields are also corrected using a running average of the normalized yields divided by the predicted yields for short holding times in order to account for any long-term drifts in the data. The normalized yields are divided by the running average to get the corrected yields.

The final lifetime value is found by doing a χ^2 minimization of the corrected yields compared to the predicted yields for each run. The statistical uncertainty on the lifetime is calculated by seeing how much the lifetime has to change in order to increase the overall χ^2 by one.

4. Results

The results from Ref. 3 of each analysis team, as well as the averages, are given in Table 1. The average lifetime is given as the average of all the analyzers' results for a particular year, and the statistical uncertainty is the average of the uncertainties from the analyzers. The current lifetime is the error-weighted average of the different years' lifetimes, and the uncertainty is the weighted standard mean. The global lifetime is similar, but includes data from 2017 and 2018. The average systematic correction to the lifetime

Table 1. Neutron lifetime results for the 2020 to 2022 UCNτ data from Ref. 3. All values are given in seconds. The four analysis teams are labeled here as A, B, C, and D. The global average includes results from 2017 and 2018, published in Ref. 5.

Analysis	A	B	C	D	Average
2020	879.47 ± 0.49	879.38 ± 0.92	879.32 ± 0.90		879.39 ± 0.89
2021	878.41 ± 0.60	878.40 ± 0.59	878.41 ± 0.55		878.41 ± 0.58
2022	876.81 ± 0.56	876.78 ± 0.58	877.06 ± 0.53	877.08 ± 0.63	876.93 ± 0.57
Current					877.96 ± 0.37
Global					877.83 ± 0.22

for the 2020 to 2022 data was 0.58 s, with a systematic uncertainty of $^{+0.20}_{-0.17}$ s.

5. Outlook

UCNτ has produced what is so far the most precise value of the neutron lifetime, with the global result for all years of data being $877.83 \pm 0.22\,(\text{stat})^{+0.20}_{-0.17}\,(\text{sys})$ s.[3] The next version of the experiment, called UCNτ+, will use a different neutron loading method. Rather than the neutrons being loaded through the bottom of the trap, a neutron "elevator" will be used to load neutrons at the top of the trap and then lower them into the trap. This new loading method is expected to increase the number of UCN in the trap by a factor of 5 to 10.[3] A new scintillator based on cerium-doped yttrium aluminum perovskite is also being developed, and should improve the uncertainty based on event definition as it should not produce the long tail of photons for neutron events that the current setup produces.[3] These improvements should allow for an even more precise measurement.

References

1. A. Czarnecki, W.J. Marciano, and A. Sirlin, *Phys. Rev. Lett.* **120**, 202002 (2018).
2. G.L. Greene and P. Geltenbort, *Sci. Am.* **314**, 36 (2016).
3. R. Musedinovic *et al.*, *Phys. Rev. C* **111**, 045501 (2025).
4. T.M. Ito *et al.*, *Phys. Rev. C* **97**, 012501 (2018).
5. F.M. Gonzalez *et al.*, *Phys. Rev. Lett.* **127**, 162501 (2021).

Chapter 24

Double Beta Decay and Lorentz Violation

River J.P. Govin

Physics Department, Indiana University, Bloomington, IN 47405, USA

We bound an a-type and multiple g-type SME coefficients in double beta decay processes. After constructing Lagrangians for both standard double beta decay and its neutrinoless counterpart, we calculate modified decay rates and half-lives for these interactions. Using collected data and estimated values for future experiments, we provide constraints on multiple coefficients that are both competitive and novel.

1. Motivation and Introduction

Double beta decay is a rare nuclear process in which a nucleus emits two electrons and two antineutrinos or, in hypothetical cases, only two electrons without any neutrinos. Within the framework of the Standard-Model Extension (SME), an effective field theory that includes violations of Lorentz and CPT symmetry,[1] double beta decay becomes a sensitive probe for new physics beyond the Standard Model. The SME introduces small Lorentz-violating coefficients that can alter the kinematics of particle processes, including the energy spectrum and angular correlations in double beta decay. These deviations can be measured in high-precision experiments, making double beta decay a valuable experimental channel for testing the fundamental symmetries of spacetime. In this work, we provide a few preliminary results.

2. Two-Neutrino Double Beta Decay

Consider the following piece of a generic gauge-theory Lagrangian incorporating Lorentz and CPT violation:

$$\mathcal{L} \supset i\bar{\psi}_A \gamma_\mu \mathcal{D}^\mu \psi_A + i^2 (a^{(5)})^{\mu\alpha\beta}_{AB} \bar{\psi}_A \gamma_\mu \mathcal{D}_\alpha \mathcal{D}_\beta \psi_B + \text{h.c.}, \tag{1}$$

where $\mathcal{D}_\mu$ is the corresponding gauge-covariant derivative, the ψ_A are spinor fields with gauge-group indices $A, B, \ldots$, and $(a^{(5)})^{\mu\alpha\beta}_{AB}$ are Lorentz-violating coefficients. We adapt this Lagrangian to the electroweak gauge symmetry to study how the isotropic, oscillation-free coefficient $(\mathring{a}^{(5)}_{\text{of}})$ affects the neutrino in a double beta decay interaction.

From the expanded Lagrangian, we are able to construct a matrix element that leads to the modified decay rate

$$d\Gamma \sim \int |\mathcal{M} + \delta\mathcal{M}(\mathring{a}^{(5)}_{\text{of}})|^2 d\Pi, \tag{2}$$

where $\mathcal{M}$ is the conventional matrix element and $\delta\mathcal{M}(\mathring{a}^{(5)}_{\text{of}})$ is the Lorentz-violating contribution. The phase space $(d\Pi)$ is also modified since the antineutrinos' dispersion relations contain $(\mathring{a}^{(5)}_{\text{of}})$.

Using data from EXO-200,[3] we are able to place an upper bound on $(\mathring{a}^{(5)}_{\text{of}})$ by finding the maximum difference between experimental values and the SME prediction. The preliminary result we calculate is

$$|\mathring{a}^{(5)}_{\text{of}}| < 218.1\,\text{GeV}^{-1}. \tag{3}$$

This should be treated as a conservative approximation for how large the possible effect is since there are many other factors that could contribute to this possible spectral distortion between data and theory.

3. Neutrinoless Double Beta Decay

Analogous to how we started in the two-neutrino case, consider a generic Lagrangian contribution of the form

$$\mathcal{L} \supset -i\psi_A g^{(6)\lambda\rho\kappa\alpha\beta}_{AB} \sigma_{\lambda\rho} D_{(\kappa} D_\alpha D_{\beta)} \psi^c_B + \text{h.c.}, \tag{4}$$

where $g^{(6)\lambda\rho\kappa\alpha\beta}_{AB}$ are Lorentz-violating coefficients. Since there are numerous components to $g^{(6)\lambda\rho\kappa\alpha\beta}_{AB}$, it is useful to construct an effective coefficient that averages over all orientations in the Sun-centered frame[6] for an estimated bound. For definiteness, we work with the following expression:

$$|g^{(6)\lambda\rho}_{\beta\beta}|^2 = \frac{1}{21}\Big(|g^{(6)\lambda\rho XXX}_{\beta\beta}|^2 + |g^{(6)\lambda\rho YYY}_{\beta\beta}|^2 + |g^{(6)\lambda\rho ZZZ}_{\beta\beta}|^2 + \cdots\Big). \tag{5}$$

Using Eq. (4), we are able to construct a modified half-life where the effective SME coefficients can be constrained by utilizing data from multiple neutrinoless double beta decay experiments.[7] Our modified half-life equation is

$$(T_{1/2}^{0\nu})^{-1} = G^{0\nu}(Z,Q)|M^{0\nu}|^2 \frac{|g_{\beta\beta}^{(6)\lambda\rho}|^2}{4R^6}, \tag{6}$$

where $G^{0\nu}(Z,Q)$ is the phase-space factor that depends on the atomic number Z of the parent nucleus and the Q-value, which is the maximum energy output by the process; $|M^{0\nu}|$ is the nuclear matrix element; and R is the atomic radius. Note that the effective SME coefficients and atomic radius replace the Majorana mass compared to the conventional half-life equation.[8]

We were able to bound these effective coefficients using data from multiple experiments and estimate bounds for future experiments. The range of values we constrained for current experiments is between 10^{-4} GeV^{-2} and 10^{-6} GeV^{-2}. For future experiments, our range becomes 10^{-6} GeV^{-2} to 10^{-8} GeV^{-2}.

We can extrapolate this process to higher mass dimensions for other g-type coefficients such as $g_{\beta\beta}^{(8)\lambda\rho\alpha\beta\kappa\zeta\eta}$ and $g_{\beta\beta}^{(10)\lambda\rho\alpha\beta\kappa\zeta\eta\xi\omega}$. Similarly, we were able to approximate limits for the effective-coefficient versions of these SME values. For current experiments, our range for $|g_{\beta\beta}^{(8)\lambda\rho}|$ and $|g_{\beta\beta}^{(10)\lambda\rho}|$ is 10^{-1} GeV^{-4} to 10^{-3} GeV^{-4} and 10^{0} GeV^{-6} to 10^{2} GeV^{-6}, respectively. For future experiments, the ranges become 10^{-3} GeV^{-4} to 10^{-5} GeV^{-4} and 10^{-2} GeV^{-6} to 10^{0} GeV^{-6}, respectively. These bounds are not just competitive, but also will be the first of their kind to be recorded in the literature.

References

1. D. Colladay and V.A. Kostelecký, *Phys. Rev. D* **58**, 116002 (1998).
2. V.A. Kostelecký and N. Russell, *Data Tables for Lorentz and CPT Violation*, 2025 edition, arXiv:0801.0287v18.
3. G. Anton *et al.* [EXO-200], *Phys. Rev. Lett.* **123**, 161802 (2019).
4. MINOS Collaboration, P. Adamson *et al.*, *Phys. Rev. Lett.* **101**, 151601 (2008); LSND Collaboration, L.B. Auerbach *et al.*, *Phys. Rev. D* **72**, 076004 (2005); J.S. Díaz *et al.*, *Phys. Rev. D* **80**, 076007 (2009).
5. F. Boehm and P. Vogel, *Physics of Massive Neutrinos*, Cambridge University Press, Cambridge, 1987.
6. R. Bluhm, V.A. Kostelecký, C.D. Lane, and N.E. Russell, *Phys. Rev. D* **68**, 125008 (2003).
7. J.S. Díaz, *Phys. Rev. D* **89**, 036002 (2014).
8. S.R. Elliott and P. Vogel, *Ann. Rev. Nucl. Part. Sci.* **52**, 115 (2002).

Chapter 25

Extending Muon-Sector Searches for Lorentz Violation with Higher-Precision Boost Transformations

Caitlyn Holl, Will Lankenau, Danny Shope, and Jay D. Tasson

Physics Department, Carleton College, Northfield, MN 55057, USA

Spin-precession observations from the Muon g-2 Experiment introduce opportunities to search for Lorentz violation. In this work, we discuss ways of using refined frame transformations to extend coverage of the Lorentz-violating parameter space in the muon sector of the minimal Standard-Model Extension, with a special focus on the Coefficient Separation Approach.

1. Introduction

The recognition of Lorentz violation as a possible indication of new physics at the Planck scale has motivated a renewed interest in tests of Lorentz symmetry.[1] Constructed as a field theory that provides an expansion about known physics in Lorentz-violating terms of increasing mass dimension, the Standard-Model Extension (SME) provides a systematic framework to search for Lorentz violation.[2] The minimal SME is made up of the terms in this expansion with the same mass dimension as the operators of known physics. In the single-fermion flat-spacetime limit, there are 35 spin-coupled Lorentz-violating degrees of freedom that are of interest in this work.

In the context of a test framework that is as broad and general as possible, the SME admits the possibility that Lorentz violation may be found associated with one type of particle and not others. Thus, muon physics offers a search opportunity that is distinct from searches with ordinary matter.[3] Relative to protons, neutrons, and electrons, fewer searches for Lorentz violation have been performed with muons, leading to a comparatively larger space of unexplored parameters.[4]

Searches for Lorentz violation in Earth-based laboratories often take advantage of the periodicities that would be generated in experiments as they move through a spacetime filled with Lorentz-violating backgrounds. The relevant motions of the lab include the boost of the Earth as it revolves around the Sun ($\beta_\oplus \sim 10^{-4}$), the boost of an Earth-based laboratory as it revolves around the rotation axis of the Earth ($\beta_L \lesssim 10^{-6}$), and the rotation of the lab at the sidereal frequency (ω). In such searches, the impact of the above boosts is typically treated perturbatively. Recent work demonstrates that the consideration of higher perturbative order in these boost transformations leads to additional sensitivities.[5,6]

There are several ways to view a search for the 35 spin-coupled degrees of freedom for the muon. One can view them one at a time in the context of a family of one-parameter models, an approach known as a Maximum-Reach Analysis, or one may effectively view them as a single 35-parameter model in the context of a Coefficient-Separation Analysis.[7] One could also consider models involving some subset greater than one and fewer than 35 of the full set of degrees of freedom. The consideration of higher-precision boost transformations has implications for both Maximum-Reach and Coefficient-Separation approaches. In this work, we focus on providing some illustrative examples of the power of higher-precision boost transformations in the context of Coefficient Separation searches.

2. The Minimal SME Muon Sector

The search for sidereal variations in muon spin-precession measurements by the Muon g-2 Collaboration is among the most sensitive searches for Lorentz violation in the muon sector, having placed the following limit on the b-háček combination of SME coefficients:[8,9]

$$\check{b}_J^{\pm} = \pm\frac{b_J}{\gamma} + md_{JT} + \tfrac{1}{2}\varepsilon_{JKL}H_{KL} \pm \frac{1}{2\gamma}m\varepsilon_{JKL}g_{KLT} \lesssim 10^{-24}\,\text{GeV}, \quad (1)$$

where $b_J, H_{KL}, d_{JT}, g_{KLT}$ are coefficients for Lorentz violation appearing in the SME action,[10] m is the mass of the muon, and γ is the relativistic factor associated with the speed of the muons relative to the lab (in the Muon g-2 Experiment, $\gamma = 29.3$). The plus sign here refers to the combination that is measurable in antimuon experiments, and the minus sign refers to the combination measurable in muon experiments. Due to the additional factor of γ, the combination of coefficients that forms $\check{b}_J$ is experiment-dependent, though in the context of a small number of measurements all involving the same speed for the muons in the lab frame, this is a convenient parameterization. In the nonrelativistic limit (as $\gamma \to 1$), the háček expressions reduce to the tilde coefficients that form a commonly used parameterization for the observable Lorentz-violating degrees of freedom in the minimal fermion sector of the SME.[4]

As we expand muon searches by considering higher-precision boost transformations, treating the observable Lorentz-violating degrees of freedom as the tilde coefficients offers an experiment-independent choice that is better suited to scenarios in which a significant fraction of the Lorentz-violating parameter space is accessible. The $\check{b}_J$ constraints can be straightforwardly translated into constraints on linear combinations of a pair of tilde coefficients as follows:

$$\check{b}_J^{\pm} = \tfrac{1}{2}(\tilde{b}_J^* - \tilde{b}_J) \pm \tfrac{1}{2\gamma}(\tilde{b}_J^* + \tilde{b}_J). \tag{2}$$

As we consider the impact of effects at higher order in the boost of the lab in the context of these searches, additional linear combinations of tilde coefficients will become accessible at lower sensitivity due to suppression by boost factors. Searches that are sensitive to different components of $\check{b}_j$ in the lab and different Fourier components of the signal can measure additional linear combinations. Moreover, some of these linear combinations are colatitude-dependent, meaning that experiments repeated at different colatitudes, as is the case for the muon g-2 experiment performed at Brookhaven[9] and Fermilab,[8] also access independent linear combinations of SME coefficients. Thus, many of these additional measurements offer added discovery potential when effects that are suppressed by the boost of the lab are considered.

3. Coefficient Separation

Combined with sufficient orders of boost expansion, suitable combinations of experimental measurements allow for full coverage of the 35-degree-of-freedom coefficient space associated with spin-coupled effects in the minimal SME for the muon. In this work, we provide a simplified example of a coefficient separation approach in the context of a small subset of tilde coefficients for Lorentz violation. Specifically, suppose that $b_X, d_{TX}, H_{YZ} = -H_{ZY}$, and $g_{YZT} = -g_{ZYT}$ are the only nonzero Sun-centered frame coefficients in our model (equalities are due to the symmetry properties of the SME coefficients). Therefore, in this limit, the only nonzero tilde coefficients are $\tilde{b}_X = b_X - H_{YZ} + mg_{YZT}$, $\tilde{b}^*_X = b_X + H_{YZ} + mg_{YZT}$, $\tilde{d}_X = md_{TX} - \frac{1}{2}H_{YZ}$, and $\tilde{g}_{DX} = -b_X + mg_{YZT}$.

As an initial example, consider the linear combinations of tilde coefficients that are accessible without suppression by laboratory boost parameters within our model. Suppose that a hypothetical experiment is capable of independently measuring periodic effects associated with $\check{b}^+_x$ and $\check{b}^+_y$ at the sidereal frequency. Further suppose that this is possible at two experimental locations with colatitudes χ_1, χ_2 respectively. We denote the experimental constraint on the linear combination associated with each of these measurements as $\mathcal{F}_{\omega,\chi_n,j}$. With this, we may write the set of four experimental measurements on linear combinations of the four nonzero tilde coefficients in our model as a matrix equation

$$\begin{bmatrix} \mathcal{F}_{\omega,\chi_1,x} \\ \mathcal{F}_{\omega,\chi_1,y} \\ \mathcal{F}_{\omega,\chi_2,x} \\ \mathcal{F}_{\omega,\chi_2,y} \end{bmatrix} = \begin{bmatrix} \frac{(1+\gamma)\cos\chi_1}{2\gamma} & \frac{(1-\gamma)\cos\chi_1}{2\gamma} & 0 & 0 \\ -\frac{2+\gamma-\gamma\cos(2\chi_1)}{4\gamma} & \frac{-2+\gamma-\gamma\cos(2\chi_1)}{4\gamma} & 0 & 0 \\ \frac{(1+\gamma)\cos\chi_2}{2\gamma} & \frac{(1-\gamma)\cos\chi_2}{2\gamma} & 0 & 0 \\ -\frac{2+\gamma-\gamma\cos(2\chi_2)}{4\gamma} & \frac{-2+\gamma-\gamma\cos(2\chi_2)}{4\gamma} & 0 & 0 \end{bmatrix} \begin{bmatrix} \tilde{b}_X \\ \tilde{b}^*_X \\ \tilde{d}_X \\ \tilde{g}_{DX} \end{bmatrix} \tag{3}$$

in which the vector of experimental measurements is equal to the product of the matrix M and the coefficient vector. As can be seen by inspection in this simple model, these measurements are insufficient to span the four-degree-of-freedom coefficient space. As an illustration of the general method that can be used in the context of the full 35-degree-of-freedom coefficient space, we can calculate the rank of M. In this example, $\text{rank}(M) = 2$, confirming that we cannot obtain four independent constraints on $\tilde{b}_X, \tilde{b}^*_X, \tilde{d}_X$, and $\tilde{g}_{DX}$ with these four experimental measurements at this level of boost suppression.

However, if we utilize higher-precision boost transformations, we can span the four-coefficient space. By considering suppression factors of up to 12 orders of magnitude, we gain sensitivity to terms in the third and fourth columns of M. Specifically, we obtain in the third column $M_{13} = -\frac{1}{2}\beta^2_\oplus \cos(\chi_1)$, $M_{23} = M_{43} = \frac{1}{2}\beta^2_\oplus + \beta^2_L$, and $M_{33} = -\frac{1}{2}\beta^2_\oplus \cos(\chi_2)$. In the fourth column, $M_{14} = \frac{1}{4\gamma}\cos(\chi_1)(\beta^2_\oplus + \beta^2_L)$, $M_{24} = M_{44} = \frac{-\beta^2_\oplus}{4\gamma}$, and $M_{34} = \frac{1}{4\gamma}\cos(\chi_2)(\beta^2_\oplus + \beta^2_L)$. Hence, with the chosen measurements, the consideration of suppressed effects increases $\text{rank}(M)$ to 4, and independent limits may be placed on the full space of tilde coefficients $\tilde{b}_X, \tilde{b}_X^*, \tilde{d}_X$, and $\tilde{g}_{DX}$.

One can also imagine a scenario in which the coefficient space is not spanned by a given set of four measurements, even at high boost suppression. Consider an example involving 12 orders of boost suppression. If we consider measurements similar to the example in Eq. (3), in which we seek Fourier components at twice the sidereal frequency rather than the sidereal frequency, we produce a similar matrix equation:

$$\begin{bmatrix} \mathcal{F}_{2\omega,\chi_1,x} \\ \mathcal{F}_{2\omega,\chi_1,y} \\ \mathcal{F}_{2\omega,\chi_2,x} \\ \mathcal{F}_{2\omega,\chi_2,y} \end{bmatrix} = \begin{bmatrix} \frac{1}{32}\beta_\oplus^2 \sin(\chi_1) & -\frac{1}{32}\beta_\oplus^2 \sin(\chi_1) & 0 & 0 \\ -\frac{1}{64}\beta_\oplus^2 \sin(2\chi_1) & \frac{1}{64}\beta_\oplus^2 \sin(2\chi_1) & 0 & 0 \\ \frac{1}{32}\beta_\oplus^2 \sin(\chi_2) & -\frac{1}{32}\beta_\oplus^2 \sin(\chi_2) & 0 & 0 \\ -\frac{1}{64}\beta_\oplus^2 \sin(2\chi_2) & \frac{1}{64}\beta_\oplus^2 \sin(2\chi_2) & 0 & 0 \end{bmatrix} \begin{bmatrix} \tilde{b}_X \\ \tilde{b}_X^* \\ \tilde{d}_X \\ \tilde{g}_{DX} \end{bmatrix}. \tag{4}$$

Note that in this case, rank$(M) = 1$. Thus, these four possible measurements are all redundant in the context of our simplified model.

Although these examples are presented in the context of a relatively small limit of the entire SME, the process carried out above is analogous to that used in a full analysis of all 35 spin-coupled degrees of freedom. Thus, work at higher powers of boost suppression forms a fruitful approach to extending searches for Lorentz violation in the muon sector.

References

1. V.A. Kostelecký and S. Samuel, *Phys. Rev. D* **39**, 683 (1989); V.A. Kostelecký and R. Potting, *Nucl. Phys. B* **359**, 545 (1991).
2. D. Colladay and V.A. Kostelecký, *Phys. Rev. D* **58**, 116002 (1998).
3. R. Bluhm, V.A. Kostelecký, and C. Lane, *Phys. Rev. Lett.* **84**, 1098 (2000).
4. V.A. Kostelecký and N. Russell, *Data Tables for Lorentz and CPT Violation,* 2025 edition, arXiv:0801.0287v18.
5. M.J. Basson *et al.*, in preparation; J.D. Tasson, these proceedings; F.M. Lopez, Z. Zhang, B.R. Lott, and J.D. Tasson, these proceedings.
6. C. Holl *et al.*, in preparation.
7. N.A. Flowers, C. Goodge, and J.D. Tasson, *Phys. Rev. Lett.* **119**, 201101 (2017).
8. M. Bhattacharya, Ph.D. Thesis, University of Mississippi, 2021.
9. G.W. Bennett *et al.*, *Phys. Rev. Lett.* **100**, 091602 (2008).
10. A. Fittante and N. Russell, *J. Phys. G* **39**, 125004 (2012).

Chapter 26

Achieving Full Coverage of the SME Minimal Matter Sector

Facundo Martin Lopez, Zhiyu Zhang, Bianca Rose Lott, and Jay D. Tasson

Physics Department, Carleton College, Northfield, MN 55057, USA

Existing experiments and data sets can be leveraged to obtain additional sensitivities to Lorentz violation, beyond those originally sought, through a more precise consideration of the boost of the experiment through the background. In fact, access to the full coefficient space of the flat-spacetime single-fermion limit of the minimal matter sector of the Standard-Model Extension can be obtained. In this work, we present this coverage for a sample particle in the context of a simplified model of Earth's motion.

1. Introduction

Lorentz violation originating from the Planck scale[1] has been intensely sought over the past quarter century[2] in the context of an effective field-theory-based framework known as the Standard-Model Extension (SME).[3,4] Searches in the minimal fermion sector with readily available ordinary matter particles (protons, neutrons, and electrons) have been among the most prevalent and sensitive. Despite this effort, the 2025 edition of the *Data Tables for Lorentz and CPT Violation*,[2] shows a number of Lorentz-violating degrees of freedom yet to be explored in experimental and observational searches. Our recent work addresses this gap by exploiting higher-precision frame transformations.[5] In this work, we summarize key aspects of our approach and provide an example of the sensitivities that are possible in the context of a simplified model of the motion of an Earth-based laboratory. We emphasize that, while the results presented here in the context of our simplified model are useful as an illustration and point of comparison with the general results, those interested in true bounds or the analysis of future experiments should refer to Ref. 5.

2. The Fermion Sector

The structure of the fermion sector is discussed in many more comprehensive resources[2,3,6] and is summarized elsewhere in these proceedings.[7] Here, we briefly define the relevant variables. The coefficients for Lorentz violation appearing in the SME Lagrange density that we explore in this study are b_μ, $d_{\mu\nu}$, $g_{\lambda\mu\nu}$, and $H_{\mu\nu}$. These coefficients result in 35 independent Lorentz-violating degrees of freedom per fermion in the minimal flat-spacetime single-fermion limit that we consider here. They are a subset of the 44 degrees of freedom per fermion typically characterized by the tilde coefficients. This subset contains the remaining unexplored degrees of freedom associated with protons, neutrons, and electrons. The explicit form of the tilde coefficients can be found in Table P56 of Ref. 2.

In what follows, we consider published results[9] based on the most sensitive class of SME experiments with fermions—those seeking the effects of the Hamiltonian[6]

$$H = \tilde{b}_j \sigma_j, \tag{1}$$

where σ_j are the Pauli spin matrices and the tilde coefficient $\tilde{b}_j$ is the combination

$$\tilde{b}_j = b_j - \tfrac{1}{2}\varepsilon_{jkl} H_{kl} - m(d_{jt} - \tfrac{1}{2}\varepsilon_{jkl} g_{klt}). \tag{2}$$

Here m is the mass of the fermion.

3. Sensitivities from Simplified Laboratory Motions

In the flat spacetime limit, the SME coefficients can be thought of as fields that are constant and uniform throughout large regions of the universe. As such, they generate time-dependent effects in the lab due to the motion of the lab through this background. A reference frame attached to the Sun is sufficiently inertial over the time scales of relevant experiments. Thus, the coefficients are taken as constants in a standard Sun-centered frame.[2] We use capital indices for components of tensors in this frame.

Coefficient	Current	Feasible	Frequency
$\tilde{b}_X$	10^{-33}	–	–
$\tilde{b}_Y$	10^{-33}	–	–
$\tilde{b}_Z$	10^{-29}	–	–
$\tilde{b}_T$	10^{-26}	$\mathbf{10^{-29}}$	$\omega_L + \Omega$
$\tilde{b}^*_X$	$\mathbf{10^{-12}}$	$\mathbf{10^{-16}}$	$\omega_L - 2\Omega$
$\tilde{b}^*_Y$	$\mathbf{10^{-15}}$	$\mathbf{10^{-16}}$	$\omega_L - 2\Omega$
$\tilde{b}^*_Z$	$\mathbf{10^{-15}}$	–	–
$\tilde{d}_+$	10^{-27}	$\mathbf{10^{-29}}$	Ω
$\tilde{d}_-$	10^{-26}	$\mathbf{10^{-29}}$	$\omega_L + \Omega$
$\tilde{d}_Q$	10^{-26}	$\mathbf{10^{-29}}$	$\omega_L - \Omega$
$\tilde{d}_{XY}$	10^{-27}	$\mathbf{10^{-29}}$	$\omega_L + \Omega$
$\tilde{d}_{YZ}$	10^{-26}	$\mathbf{10^{-28}}$	$\omega_L + \Omega$
$\tilde{d}_{ZX}$	$\mathbf{10^{-27}}$	$\mathbf{10^{-29}}$	Ω
$\tilde{d}_X$	10^{-28}	–	–
$\tilde{d}_Y$	10^{-28}	–	–
$\tilde{d}_Z$	$\mathbf{10^{-24}}$	–	–
$\tilde{H}_{XT}$	$\mathbf{10^{-27}}$	$\mathbf{10^{-29}}$	Ω
$\tilde{H}_{YT}$	$\mathbf{10^{-27}}$	$\mathbf{10^{-29}}$	Ω
$\tilde{H}_{ZT}$	10^{-27}	$\mathbf{10^{-29}}$	$\omega_L - \Omega$

Sensitive experiments seeking the periodic effects associated with motions of the Earth-based laboratories are key tools in searching for Lorentz violation. In this work, we consider the most significant of these motions: the boost of the Earth as it orbits the Sun ($\beta_\oplus \sim 10^{-4}$) at the annual frequency Ω, which we take as a fixed circular orbit; the rotation of the Earth at the sidereal frequency ω, which we take as occurring around a fixed axis; and the boost of the laboratory as it revolves around the Earth ($\beta_L \lesssim 10^{-6}$) at the sidereal frequency, which we explore in the context of a perfect-sphere Earth approximation. We treat the relevant frame transformations, which are described in detail elsewhere in these proceedings,[7] perturbatively in the small quantities $\beta_\oplus$ and β_L. We proceed to sufficiently high order in these quantities to obtain sensitivities to all 35 remaining degrees of freedom in this limit of the SME, whereas few prior SME searches have considered boost effects beyond linear order in such a perturbation series.

We first consider a Maximum Reach Analysis,[8] which places constraints on Lorentz violation by assuming one nonzero Lorentz-violating degree of freedom at a time. The sensitivities that result from a Maximum Reach Analysis[8] that uses this model of laboratory motion and published bounds from neutron experiments[9] are found in Tables 1 and 2. A Coefficient Separation Approach[8] considers a family of coefficients together. Consideration of sufficiently high order in our perturbative expansion offers exciting new opportunities for Coefficient Separation searches as well.[5,10]

Coefficient	Current	Feasible	Frequency
$\tilde{g}_T$	10^{-27}	$\mathbf{10^{-29}}$	$\omega_L + \Omega$
$\tilde{g}_c$	10^{-27}	$\mathbf{10^{-29}}$	$\omega_L + \Omega$
$\tilde{g}_Q$	$\boxed{\mathbf{10^{-24}}}$	–	–
$\tilde{g}_-$	$\boxed{\mathbf{10^{-24}}}$	–	–
$\tilde{g}_{TX}$	$\boxed{\mathbf{10^{-24}}}$	–	–
$\tilde{g}_{TY}$	$\boxed{\mathbf{10^{-24}}}$	–	–
$\tilde{g}_{TZ}$	$\boxed{\mathbf{10^{-24}}}$	–	–
$\tilde{g}_{XY}$	$\boxed{\mathbf{10^{-19}}}$	$\boxed{\mathbf{10^{-20}}}$	Ω
$\tilde{g}_{YX}$	$\boxed{\mathbf{10^{-19}}}$	$\boxed{\mathbf{10^{-20}}}$	Ω
$\tilde{g}_{ZX}$	$\boxed{\mathbf{10^{-18}}}$	$\boxed{\mathbf{10^{-20}}}$	Ω
$\tilde{g}_{XZ}$	$\boxed{\mathbf{10^{-18}}}$	$\boxed{\mathbf{10^{-20}}}$	Ω
$\tilde{g}_{YZ}$	$\boxed{\mathbf{10^{-18}}}$	$\boxed{\mathbf{10^{-20}}}$	$\omega_L - 3\Omega$
$\tilde{g}_{ZY}$	$\boxed{\mathbf{10^{-18}}}$	$\boxed{\mathbf{10^{-20}}}$	Ω
$\tilde{g}_{DX}$	10^{-28}	–	–
$\tilde{g}_{DY}$	10^{-28}	–	–
$\tilde{g}_{DZ}$	$\boxed{\mathbf{10^{-24}}}$	–	–

4. Discussion

The results of our analysis of higher-order boost effects in the simplified laboratory-motion approximation generate a useful point of comparison with the results of our more comprehensive treatment.[5] We observe that within the simplified model, like the more comprehensive treatment, full coverage of the remaining degrees of freedom in the minimal fermion sector is possible in a Maximum Reach Analysis using the published results of experiments sensitive to the interaction in Eq. (1). Our more comprehensive treatment results in relatively minor refinements on the basic model used here.[5]

The necessary calculations to obtain the results in Tables 1 and 2 are quite involved. To facilitate a greater intuition for how these results come about, some examples obtained in the context of special limits of the full coefficient space are presented elsewhere in these proceedings.[7] In our comprehensive results,[5] we consider the electron and proton sector in addition to the neutron sector. We also consider the muon sector in other work.[10,11] The results obtained here in the context of the simplified model of laboratory motion demonstrate the potential of considering higher-order effects from the boost of the Earth in extending the sensitivities of experimental searches for Lorentz violation.

References

1. V.A. Kostelecký and S. Samuel, *Phys. Rev. D* **39**, 683 (1989).
2. V.A. Kostelecký and N. Russell, *Data Tables for Lorentz and CPT Violation,* 2025 edition, arXiv:0801.0287v18.
3. D. Colladay and V.A. Kostelecký, *Phys. Rev. D* **58**, 116002 (1998).
4. V.A. Kostelecký, *Phys. Rev. D* **69**, 105009 (2004).
5. M.J. Basson *et al.*, in preparation.
6. V.A. Kostelecký and C.D. Lane, *J. Math. Phys.* **40**, 6245 (1999).
7. J.D. Tasson, these proceedings.
8. N.A. Flowers, C. Goodge, and J.D. Tasson, *Phys. Rev. Lett.* **119**, 201101 (2017).
9. F. Allmendinger *et al.*, *Phys. Rev. Lett.* **112**, 110801 (2014).
10. C. Holl *et al.*, in preparation.
11. C. Holl *et al.*, these proceedings.

Chapter 27

Nuclear $\mu - e$ Conversion via Lorentz and CPT Violation

William P. McNulty

Department of Physics, Indiana University, Bloomington, IN 47408, USA

One of the three common experimental channels searching for charged lepton flavor violation (CLFV) via $\mu - e$ conversion involves first capturing the muon on a nuclear target. This channel provides a unique opportunity to constrain four-point quark-lepton operators that may contribute to CLFV, in addition to the electromagnetic operators that can also be accessed in the other channels. We investigate the leading Lorentz- and CPT-violating contributions for nuclear $\mu - e$ conversion experiments within the Standard-Model Extension and obtain the first bounds on the relevant quark-lepton operators using the results of the SINDRUM II experiment; we also consider possibilities for improved constraints from upcoming experiments.

1. Background

Lepton flavor conservation is an accidental symmetry of the Standard Model (SM), arising as a consequence of leaving the neutrinos massless. The existence of neutrino oscillations due to the nonvanishing masses of at least two neutrino species manifestly violates lepton flavor conservation, and implies that charged lepton flavor violation (CLFV) should also occur. However, without extending the SM beyond the inclusion of massive neutrinos, such processes are expected to be highly suppressed by the smallness of the neutrino masses and the GIM mechanism, with branching ratios $\lesssim 10^{-54}$.[1] Since this is well beyond current sensitivities, any observation of CLFV would be interpreted as a signal of new physics beyond the SM; thus, such processes have been the subject of an intense program of experimental searches.

In the muon-electron sector of CLFV searches, one of the so-called "golden channels" is the coherent conversion of a muon to an electron in the presence of a nucleus, $\mu^- + {}^A_Z N \to e^- + {}^A_Z N$; here, "coherent" means that the nucleus remains in its ground state throughout the interaction. The experiment is performed by colliding a muon beam with an atomic target, on which some of the muons will capture to form short-lived muonic atoms, with the muon quickly falling to the 1S ground state. The majority of these muons will be captured by the nucleus through the exchange of a W boson: $\mu^- + {}^A_Z N \to \nu_\mu + {}^{A}_{Z-1}N$. The remainder are expected to decay in orbit, $\mu^- \to e^- \overline{\nu_e} \nu_\mu$, with the resulting electron exhibiting the energy spectrum corresponding to a three-body decay. The goal of the experiment, however, is to observe the CLFV process $\mu^- + {}^A_Z N \to e^- + {}^A_Z N$. This produces a monoenergetic electron with energy $E_e^{\text{conv}} = m_\mu - E_{\text{bind}} - E_{\text{recoil}}$, where m_μ is the muon mass, E_{bind} is the binding energy of the muonic atom in the 1S state, and E_{recoil} is the (negligible) nuclear recoil energy.

To date, of course, this process is not known to have been observed; therefore, experiments set upper bounds on the ratio of the rate of the CLFV process to the capture rate:

$$R_{\mu e} \equiv \frac{\omega_{\text{conv}}}{\omega_{\text{capt}}} = \frac{\Gamma(\mu^- + {}^A_Z N \to e^- + {}^A_Z N)}{\Gamma(\mu^- + {}^A_Z N \to \nu_\mu + {}^{A}_{Z-1}N)}. \tag{1}$$

The current best limit on this ratio has been set by the SINDRUM II experiment at the Paul Scherrer Institute (PSI): $R_{\mu e} < 7 \times 10^{-13}$ at 90% confidence level (CL) on a gold (${}^{197}_{79}\text{Au}$) target.[2] The next generation of experiments consists of COMET at the Japan Proton Accelerator Complex (J-PARC) and Mu2e at the Fermi National Accelerator Laboratory (Fermilab), both on aluminum (${}^{27}_{13}\text{Al}$) targets. COMET will target a sensitivity of $R_{\mu e} \simeq 7 \times 10^{-15}$ in Phase I, with a final goal of $R_{\mu e} \simeq 2.6 \times 10^{-17}$, both at 90% CL.[3] Mu2e expects a sensitivity of $R_{\mu e} \simeq 6.2 \times 10^{-16}$ from Run I, with a final goal of $R_{\mu e} \simeq 8 \times 10^{-17}$, both at 90% CL.[4]

2. Conversion Rate in the Standard-Model Extension

The Standard-Model Extension (SME) is the effective field theory that expands on the SM by incorporating potential violations of Lorentz invariance and the CPT symmetry. The leading contributions to nuclear $\mu - e$

conversion in the SME are expected to come from electromagnetic operators of mass dimension five and four-point quark-lepton operators of mass dimension six. At tree level, the quark-lepton operators are uniquely accessed in this experimental channel, while the electromagnetic operators can also be constrained by searches in the other "golden channels."

To account for the presence of the nucleus, one first needs to obtain the leptonic wavefunction, which can be written

$$\psi_{\kappa,s}(r,\theta,\phi) = \begin{pmatrix} g(r)\chi_{\kappa,s}(\theta,\phi) \\ if(r)\chi_{-\kappa,s}(\theta,\phi) \end{pmatrix}. \tag{2}$$

The eigenspinors satisfy $(\boldsymbol{\sigma}\cdot\mathbf{l}+I)\chi_{\kappa,s} = -\kappa\chi_{\kappa,s}$ and $J_z\chi_{\kappa,s} = s\chi_{\kappa,s}$ for $\kappa = \mp(J+1/2)$. The radial wavefunctions are obtained by numerically solving the Dirac equation using spherically-symmetric nuclear charge distributions.[5,6]

The conversion rate for coherent conversion via the electromagnetic operators can be written as

$$\omega^{(5)}_{\text{conv}} = \frac{1}{2}\sum_{s=\pm\frac{1}{2}}\sum_{\kappa=\pm 1}\sum_{s'=\pm\frac{1}{2}}\left|\int d^3x\ \bar{\psi}^{(e)}_{\kappa,s'}\mathcal{O}^{\alpha\beta}\psi^{(\mu)}_s F_{\alpha\beta}\right|^2, \tag{3}$$

where $F_{\alpha\beta}$ is the electromagnetic field strength tensor containing the electric field of the nucleus, $\bar{\psi}^{(e)}_{\kappa s'}$ is the continuum wavefunction of an electron of energy E^{conv}_e and spin s', and $\psi^{(\mu)}_s$ is the 1S bound-state wavefunction of a muon of spin s. The SME operators are denoted by $\mathcal{O}^{\alpha\beta} \in \{(m^{(5)}_F)^{\alpha\beta}_{\mu e}, i(m^{(5)}_{5F})^{\alpha\beta}_{\mu e}\gamma^5, (a^{(5)}_F)^{\mu\alpha\beta}_{\mu e}\gamma_\mu, (b^{(5)}_F)^{\mu\alpha\beta}_{\mu e}\gamma^5\gamma_\mu, \frac{1}{2}(H^{(5)}_F)^{\mu\nu\alpha\beta}_{\mu e}\sigma_{\mu\nu}\}$, with the restriction that one index is temporal and the other spatial (as only the electric field is considered).

The conversion rate via the quark-lepton operators can be written as

$$\omega^{(6)}_{\text{conv}} = \frac{1}{2}\sum_{s=\pm\frac{1}{2}}\sum_{\kappa=\pm 1}\sum_{s'=\pm\frac{1}{2}}\left|\int d^3x\left(\alpha\bar{\psi}^{(e)}_{\kappa,s'}\mathcal{K}\psi^{(\mu)}_s + \beta\bar{\psi}^{(e)}_{\kappa,s'}\mathcal{K}_0\psi^{(\mu)}_s\right)\right|^2, \tag{4}$$

where the nuclear matrix elements that are nonvanishing for coherent conversion[6] are denoted by $\alpha = \sum_q \langle N|\bar{\psi}^{(q)}\psi^{(q)}|N\rangle$ for $q \in \{u,d,s\}$ and $\beta = \sum_q \langle N|\bar{\psi}^{(q)}\gamma_0\psi^{(q)}|N\rangle$ for $q \in \{u,d\}$. The SME operators are denoted by $\mathcal{K} \in \{(k^{(6)}_{SV})^{\lambda}_{qq\mu e}\gamma_\lambda, (k^{(6)}_{SA})^{\lambda}_{qq\mu e}\gamma^5\gamma_\lambda, \frac{1}{2}(k^{(6)}_{ST})^{\kappa\lambda}_{qq\mu e}\sigma_{\kappa\lambda}\}$ and $\mathcal{K}_0 \in \{(k^{(6)}_{VS})^{t}_{qq\mu e}, i(k^{(6)}_{VP})^{t}_{qq\mu e}\gamma^5, \frac{1}{2}(k^{(6)}_{VV})^{t\lambda}_{qq\mu e}\gamma_\lambda, (k^{(6)}_{VA})^{t\lambda}_{qq\mu e}\gamma^5\gamma_\lambda, \frac{1}{2}(k^{(6)}_{VT})^{t\kappa\lambda}_{qq\mu e}\sigma_{\kappa\lambda}\}$.

Although initial calculations may be done in the lab frame in spherical coordinates, ultimately the results must be transformed to the

Table 1. Constraints based on SINDRUM II results.

Operator Class	Constraint
Electromagnetic	$< (6\text{–}8) \times 10^{-12}\,\text{GeV}^{-1}$
Scalar Quark-Lepton (u, d)	$< (6\text{–}7) \times 10^{-13}\,\text{GeV}^{-2}$
Scalar Quark-Lepton (s)	$< (10\text{–}15) \times 10^{-13}\,\text{GeV}^{-2}$
γ_0 Quark-Lepton (u, d)	$< (20\text{–}30) \times 10^{-13}\,\text{GeV}^{-2}$

Sun-Centered Frame (SCF). The SCF is defined such that $T = 0$ corresponds to the 2000 vernal equinox, the X axis points from the Earth to the Sun at that time, and the Z axis is parallel to the rotation axis of the Earth.[7] The transformation to a standard laboratory frame (x axis towards local south, y axis towards local east) on Earth is given in Eq. (7) of Ref. 8 and in general depends on the colatitude of the laboratory, the sidereal frequency of the Earth, and the sidereal time (shifted from T by an amount that depends on longitude, among other effects).[9] The experiments considered here are taken to have 2π azimuthal coverage. Acceptance in the polar angle is restricted to a region symmetric about $\theta = \pi/2$ with respect to the beamline, which defines the polar z axis; this is related to the z axis in the standard lab frame by an additional improper rotation.

3. Results and Outlook

We obtain constraints on the relevant SME operators using the results of the SINDRUM II experiment. A summary of these results is provided in Table 1; detailed results, along with projections for the initial phases of COMET and Mu2e, are available in Ref. 10. In general, we expect to see at least an order of magnitude improvement due to the initial phases of the upcoming experiments, while their final sensitivity goals suggest they could ultimately provide two orders of magnitude improvement.

The constraints on the electromagnetic operators coming from the SINDRUM II results are weaker, by about an order of magnitude, than those previously obtained[8] using the results of the MEG experiment at PSI, which probed $\mu^+ \to e^+\gamma$.[11] The MEG II experiment is currently taking data[12] and would be expected to further strengthen those bounds, although the later-stage results of COMET and Mu2e would likely produce competitive bounds for these operators.

The constraints on the four-point quark-lepton operators are the first reported[13] for these SME coefficients, which are experimentally accessible at tree level only in the nuclear conversion channel; this channel thus remains complementary to the other $\mu - e$ conversion searches. These constraints are expected to be improved by an order of magnitude or more from just the initial phases of the upcoming COMET and Mu2e experiments.

Acknowledgments

This contribution to the proceedings is based on Ref. 10, in collaboration with V. Alan Kostelecký, Emilie Passemar, and Nathaniel Sherrill. This contribution is supported in part by the U.S. National Science Foundation under grant PHY-2013184.

References

1. W.J. Marciano and A.I. Sanda, *Phys. Lett. B* **67**, 303 (1977); S.M. Bilenky, S.T. Petcov, and B. Pontecorvo, *Phys. Lett. B* **67**, 309 (1977); B.W. Lee, S. Pakvasa, R.E. Schrock, and H. Sugawara, *Phys. Rev. Lett.* **38**, 937 (1977).
2. A. van der Schaaf, *J. Phys. G: Nucl. Part. Phys.* **29**, 1503 (2003).
3. COMET Collaboration, R. Abramishvili *et al.*, *PTEP* **2020**, 033C01 (2020).
4. Mu2e Collaboration, K. Byrum *et al.*, *Universe* **2023**, *9*, 54 (2023); R.H. Bernstein, *Front. Phys.* **27**, 1 (2019).
5. H. de Vries, C.W. de Jager, and C. de Vries, At. *Data Nucl. Data Tables* **36**, 495 (1987).
6. R. Kitano, M. Koike, and Y. Okada, *Phys. Rev. D* **66**, 096002 (2002).
7. R. Bluhm, V.A. Kostelecký, C.D. Lane, and N. Russell, *Phys. Rev. D* **68**, 125008 (2003); *Phys. Rev. Lett.* **88**, 090801 (2002); V.A. Kostelecký and M. Mewes, *Phys. Rev. D* **66**, 056005 (2002).
8. V.A. Kostelecký, E. Passemar, and N. Sherrill, *Phys. Rev. D* **106**, 076016 (2022).
9. V.A. Kostelecký, A.C. Melissinos, and M. Mewes, *Phys. Lett. B* **761**, 1 (2016); Y. Ding and V.A. Kostelecký, *Phys. Rev. D* **94**, 056008 (2016).
10. V.A. Kostelecký, W.P. McNulty, E. Passemar, and N. Sherrill, *Phys. Lett. B* **868**, 139755 (2025).
11. MEG Collaboration, A.M. Baldini *et al.*, *Eur. Phys. J. C* **76**, 434 (2016).
12. MEG II Collaboration, K. Afanaciev *et al.*, *Eur. Phys. J. C* **84**, 216 (2024).
13. V.A. Kostelecký and N. Russell, *Data Tables for Lorentz and CPT Violation,* 2025 edition, arXiv:0801.0287v18.

Chapter 28

Dalitz Plot Kinematics for a Lorentz-Violating Three-Body Decay

Joshua O'Connor

Department of Physics and Astronomy, University of South Carolina, Columbia, South Carolina 29208, USA

We outline the analysis of three-particle interaction and decay processes for leading-order modifications in a Lorentz-violating quantum field theory.

1. Introduction

Modern particle theories typically emphasize calculations and modifications thereof to the transition matrix. When considering the inclusion of symmetry-breaking effects, however, the kinematic modifications made in formal scattering theory can be close to, or in many cases even more important than, changes to matrix elements. In physical theories, the estimated available phase space can produce useful results; for example, in Ref. 1 the exact radiation rate for a Cerenkov-type effect with modified field equations was found to agree well with the kinematically allowed region estimate.[1]

We shall define the SME theory that is applied in our kinematics calculation. Practicalities of three-body decays are given using a description of the Lorentz-violating background tensor $c_{\mu\nu}$ in the modified phase space. Then we give an overview of the kinematics in terms of Dalitz plots and propose possibilities for extensions of this type of prescription.

2. Lorentz-Violating Field Theory in Phase Space

Our model is developed in the fermionic sector of the Standard-Model Extension (SME). With the use of the spin-independent coefficients $c_{\mu\nu}$ (a symmetric two-index tensor background), the results are equivalent to those in a scalar sector, since our procedure is based purely on modified phase space. In Refs. 2, 3, the basis for any general local field theory that involves Lorentz-violating modifications to physical processes containing known Standard-Model fields is derived. The $c_{\mu\nu}$ coefficients describe violations of rotation and boost symmetries, but without any breaking of CPT.

Consider an η meson or kaon as the Lorentz-invariant initial particle, which then decays in the center-of-mass frame into three Lorentz-violating spinless particles of equal mass, such as pions. A compact description of the kinematics is contained in the dispersion relation given for the pion field:

$$(g^{\mu\nu} + 2c^{\mu\nu})\, p_\mu p_\nu - m^2 = 0. \tag{1}$$

The Lorentz-noninvariant modification to the kinematics allows us to calculate the three-particle decay rate with modified particle momenta and energies.[4] The main calculation of the decay rate is modified, as opposed to the original Lorentz-symmetric case, by the inclusion in the integrand of modified energy-momentum-conserving Dirac δ-functions. These dispersion relations, kinematics, and further details are laid out in Ref. 5, the manuscript which this contribution outlines. The leading-order decay rate that results without neglecting any coefficients under a particular texture of $c_{\mu\nu}$ modifications simplifies to

$$\Gamma = \frac{|\mathcal{M}|^2}{512\pi^2\sqrt{s}}\left[s(1+2c) + 4cm^2 + 4cm^2 \log\left(\frac{2mc}{\sqrt{s}-2m}\right)\right]. \tag{2}$$

The last term in the expression is recognized as the Lorentz-noninvariant analog of an ultraviolet divergence, with poles at the correct values of the particle mass. Regulation techniques and asymptotic approximations could be further applied to models of this type.[6] The leading-order series expansion results in the above expression for the decay rate, which depends not only on first-order $c_{\mu\nu}$ coefficients but also on a larger additional logarithmic enhancement term of $\mathcal{O}(c \log c)$.

We can more fully express the kinematics using modified Dalitz-plot boundaries. In this case, the elongation of the curve depends only on the available decay energy compared with the common mass of the daughters, $\sqrt{s}/m$. If $c_{\mu\nu}$ is nonzero, the boundary plot shifts according to the dashed and dotted curves in each plot. This is linked to the fact that, under Lorentz transformation, the allowed physical region will be rotated and shifted relative to the case without Lorentz violation. Figure 1 shows the effect of $c_{\mu\nu}$ on the shape of the Dalitz plot for a decay with $\sqrt{s}/m = 4.0$, for the physical decay $\eta \to 3\pi^0$. Precision measurements of this particular decay are already recognized to be important, since the matrix element, calculated using chiral perturbation theory, depends directly on the isospin-violating up–down mass difference, $m_d - m_u$.[7–9] The higher energy and ultrarelativistic plots are shown in Figs. 2 and 3. Lorentz violation changes the shape of the outline for three-pion decays, with more pronounced fractional changes at lower values of $\sqrt{s}/m$.

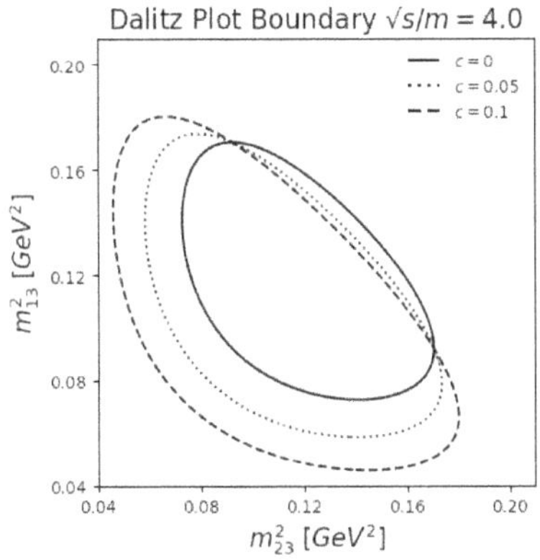

Fig. 1. Shape of the Dalitz plot for the $\sqrt{s}/m$ value corresponding to $\eta \to 3\pi^0$. The standard outline is shown, along with outlines for two nonzero values of the Lorentz-violation parameter $c_{\mu\nu}$.

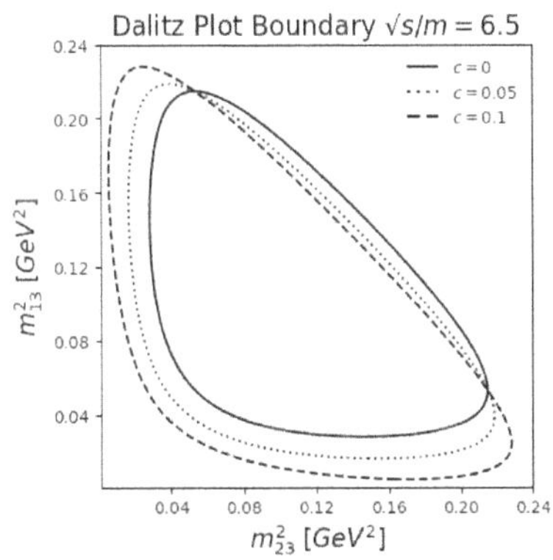

Fig. 2. Shape of the Dalitz plot for a $\sqrt{s}/m = 6.5$ value.

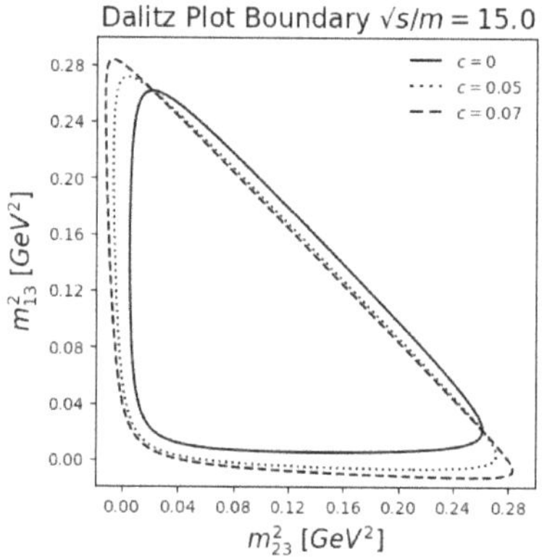

Fig. 3. Shape of the Dalitz plot for an ultrarelativistic value of $\sqrt{s}/m = 15.0$.

Increasing the strength of the Lorentz violation (in this specific model), shifts the kinematically allowed region toward smaller values of m_{23}^2 and m_{13}^2—in a symmetric fashion, because the daughter particles are identical. In the strongly relativistic regime, as the shape becomes increasingly triangular, we can see that it correspondingly becomes relatively less sensitive to the variety of Lorentz violation that we have considered. This appears to be a generic feature of $c_{\mu\nu}$-type Lorentz violation. This is perhaps not unexpected, with a $c_{\mu\nu}$ coefficient tending to shift the boundaries of kinematically allowed regions by less as the pion masses become less important and the energy-momentum relations for the outgoing particles become nearly linear.

3. Conclusions and Outlook

Calculations in the presence of Lorentz violation increase the number of singularities in the decay integral and boundary equation. The anisotropy of Lorentz violation is partially responsible for these complications, since the particle energies now depend on the angular orientations of the daughter particles. In our model, whether or not in the presence of nonzero Lorentz-symmetry-breaking coefficients, our plot is symmetric, since forming the scalar quantities m_{13}^2 and m_{23}^2 drops the leading-order directional information. If this type of analysis were applied to weak-sector β decays, such as muon decay,[10,11] where the daughter particles are not identical scalars, the convex outline could take a more asymmetrical shape. Phase-space modifications of this type could be generalized to a broad class of SME coefficients, such as the charged-fermion sectors.[12,13]

More elaborate versions of the model used here, with more complicated $c_{\mu\nu}$ textures, could be of use along with time-dependent pion data sets to provide signatures or bounds for Lorentz violation. The Dalitz plots can be binned for different sidereal times to find shifts in the physically allowed region as the laboratory rotates with the Earth. There have been some previous considerations of how reaction thresholds for particle processes would change with the Earth's rotation, but the Dalitz plot provides a more general structure.

Acknowledgments

The author thanks both B. Altschul and M. Schindler for working with him and for their time in very useful discussions.

References

1. R. Lehnert, R. Potting, *Phys. Rev. D* **70**, 125010 (2004); erratum *Ibid.* **70**, 129906 (2004).
2. D. Colladay, V.A. Kostelecký, *Phys. Rev. D* **55**, 6760 (1997).
3. D. Colladay, V.A. Kostelecký, *Phys. Rev. D* **58**, 116002 (1998).
4. B. Altschul, D. Colladay, *Phys. Rev. D* **71**, 125015 (2005).
5. J. O'Connor, B. Altschul, submitted to *Phys. Rev. D.*
6. A. Alonso Izquierdo, W. Garcia Fuertes, M.A. Gonzalez Leon, J. Mateos Guilarte, *Nucl. Phys. B* **635**, 525 (2002).
7. A.V. Anisovich, H. Leutwyler, *Phys. Lett. B* **375**, 335 (1996).
8. J. Bijnens, K. Ghorbani, *JHEP* **11**, 030 (2007).
9. K. Kampf, M. Knecht, J. Novotný, M. Zdráhal, *Phys. Rev. D* **84**, 114015 (2011).
10. K.K. Vos, H.W. Wilschut, R.G.E. Timmermans, *Rev. Mod. Phys.* **87**, 1483 (2015).
11. A. Sytema *et al.*, *Phys. Rev. C* **94**, 025503 (2016).
12. B. Altschul, *Phys. Rev. D* **75**, 105003 (2007).
13. M. Schreck, *Phys. Rev. D* **96**, 095026 (2017).

Chapter 29

Constraining e^- Lorentz Violation Through Compton Scattering

Connor A. Petway

Department of Physics, Indiana University, Bloomington, IN 47405, USA

Presented here is a method to obtain experimental bounds on electron-sector SME coefficients by utilizing data from Compton polarimeters. These polarimeters have been and will be used in many present and future experiments involving polarized-electron scattering. This allows for the possibility of obtaining better experimental bounds in the future.

1. Motivation

There are currently only five coefficients in the minimal electron sector of the SME that remain unbounded: $\tilde{g}_Q$, $\tilde{g}_-$, $\tilde{g}_{TX}$, $\tilde{g}_{TY}$, and $\tilde{g}_{TZ}$.[1] I am proposing to use Compton polarimetry to obtain bounds on these coefficients. Analyses of Compton scattering with Lorentz violation have already proven useful in constraining SME coefficients in the photon sector,[2–4] and similarly, this process is sensitive to effects from the electron sector. The utility of these analyses comes from the high sensitivity to slight changes in the kinematics of this process at high beam energies. Future polarimeters intend to operate at even higher beam energies than current experiments, promising more sensitivity to Lorentz violation.[5]

2. Modified Dispersion

To obtain bounds on the five unconstrained coefficients mentioned in Section 1, we must determine the modified energy eigenvalues for the electron considering only these coefficients. The electron Lagrangian of interest is therefore given by

$$\mathcal{L} = \bar{\psi}\left[i\left(\gamma^\mu + \tfrac{1}{2}g^{\kappa\lambda\mu}\sigma_{\kappa\lambda}\right)\partial_\mu - m_e\right]\psi. \tag{1}$$

The inclusion of this additional term produces a correction to the electron dispersion relation:

$$p^2 - m_e^2 \approx \pm 2\sqrt{-g_{\mu\alpha\beta}g^\mu{}_{\rho\sigma}p^\alpha p^\beta p^\rho p^\sigma}. \tag{2}$$

The experiments of interest typically use ultra-relativistic electrons. In this high-energy limit, the Lorentz-violating effects can be encapsulated by the definition of a parameter $\varepsilon(\hat{p})$ such that

$$E(\vec{p}) = E_0(p)(1 \pm \varepsilon(\hat{p})), \tag{3}$$

where $E_0(p) = \sqrt{p^2 + m_e^2}$ and $\varepsilon(\hat{p}) = \sqrt{-g_{\mu\alpha\beta}g^\mu{}_{\rho\sigma}\hat{p}^\alpha\hat{p}^\beta\hat{p}^\rho\hat{p}^\sigma}$. Here, we use the notation $\hat{p}^\mu = (1, \hat{p})$.

3. Compton Polarimetry

Compton scattering is often used to measure electron beam polarization. Here we describe the process used by the Compton polarimeter for the Q_{Weak} experiment at Jefferson Lab's Hall C pictured in Fig. 1.[6,7] However, the process is similar in other polarimeters, and an analogous analysis will be possible for their data as well. In this process, electrons are deflected by a dipole magnet off the main beamline and into an optical cavity, where they are then scattered off polarized photons. The unscattered electrons are deflected back into the main beamline. Having lost some of their energy to the photons, the scattered electrons are deflected farther into a diamond microstrip detector. The asymmetry between left- and right-handed scattering counts is used to measure the electron polarization P_e. This is done by taking the experimental asymmetry factor

$$A_{\text{exp}} = \frac{Y^+ - Y^-}{Y^+ + Y^-}, \tag{4}$$

where $Y^\pm$ are the charge-normalized yields in the detectors, and comparing it to the theoretical expression

$$A_{\text{th}} = \frac{\sigma^+ - \sigma^-}{\sigma^+ + \sigma^-}, \tag{5}$$

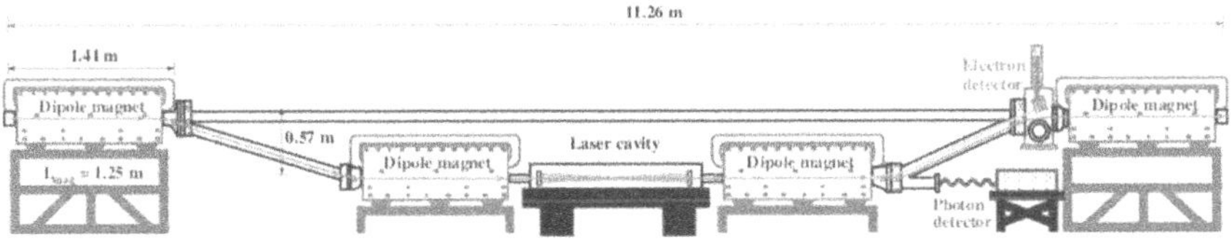

Fig. 1. Schematic of the Hall C polarimeter at Jefferson Lab. Figure taken from the technical documentation of the apparatus.[6]

where $\sigma^{\pm}$ are the theoretical cross sections, using the following equation:[6]

$$A_{\rm exp} = P_e P_\gamma A_{\rm th}. \tag{6}$$

Both $A_{\rm exp}$ and $A_{\rm th}$ are functions of $\rho = \omega/\omega_{\rm max}$, the outgoing photon energy normalized to the maximum value $\omega_{\rm max}$. While differences between $A_{\rm exp}$ and $A_{\rm th}$ can indicate decreased beam polarization, they can also be a sign of new physics not considered in the calculation of the theoretical asymmetry.

4. Lorentz-Violating Correction to $A_{\rm th}$

The modification to the dispersion relation shown in Section 2 produces a modification to the kinematics that affects $A_{\rm th}$:

$$A_{\rm th,LV} = A_{\rm th,LI} + \delta A_{\rm th}, \tag{7}$$

where

$$\delta A_{\rm th} = 4\gamma^2 f(\rho, a)\varepsilon(\hat{p}). \tag{8}$$

Here, γ is the incoming electron's Lorentz factor, and $a = 4\gamma\omega_0 \sin^2(\frac{\theta_0}{2})/m_e$ characterizes the incoming photon energy, ω_0, and the crossing angle θ_0. The SM and modified theoretical asymmetries are compared in Fig. 2 and $\delta A_{\rm th}$ itself is plotted in Fig. 3. The factor of γ^2 is responsible for the enhanced sensitivity to the Lorentz-violating effects from the kinematics as opposed to all other effects. The function $f(\rho, a)$ has been determined by calculating the Compton asymmetry for each energy-eigenstate transition and averaging over the these values.

Since $\delta A_{\rm th}$ is also a function of ρ, the effects of Lorentz violation are easily distinguished from effects of decreased beam polarization. After determining P_e, one can fit $(A_{\rm exp} - P_e P_\gamma A_{\rm th,LI})$ to $\delta A_{\rm th}$ to obtain a measurement of $\varepsilon(\hat{p})$.

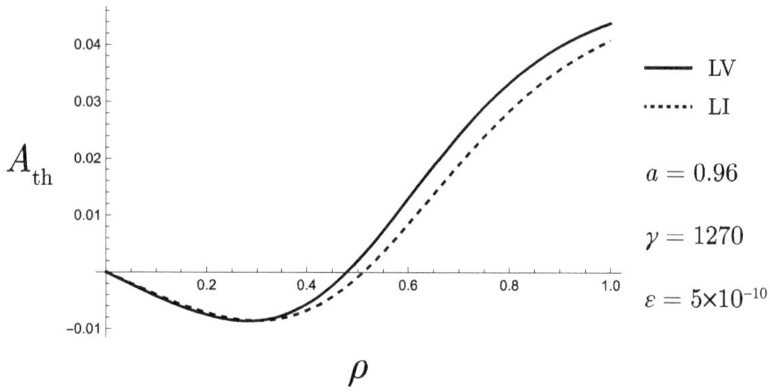

Fig. 2. Graph of both the Lorentz-invariant (LI) and Lorentz-violating (LV) theoretical Compton asymmetries, A_{th}, for kinematic parameters similar to those of Jefferson Lab's Hall C polarimeter[6,7] and an arbitrary value of ε.

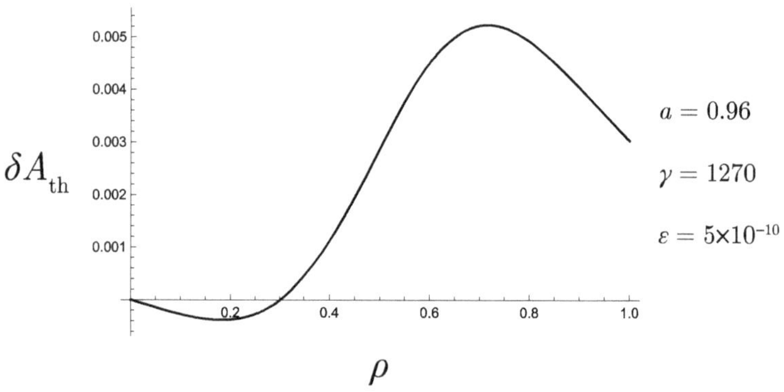

Fig. 3. Graph of the LV correction to the theoretical Compton asymmetry, δA_{th}, again using kinematic parameters similar to those of Jefferson Lab's Hall C polarimeter[6,7] and an arbitrary value of ε.

5. Sidereal Variations

Since $\varepsilon(\hat{p})$ depends on the electron momentum direction, it will experience harmonic variations at the sidereal frequency. For simplicity of the expressions, it is easier to work with ε^2 than ε. Because ε^2 is fourth order in $\hat{p}$, it should produce variations up to the fourth harmonic. It can therefore be expressed as a finite Fourier series:

$$\varepsilon^2 = B_0 + \sum_{n=1}^{4} [A_n \sin(n\omega_\oplus T_\oplus) + B_n \cos(n\omega_\oplus T_\oplus)]. \tag{9}$$

These Fourier coefficients can be constrained by plotting ε^2 over time and analyzing its variations. The bounds on these Fourier coefficients can then be used to calculate constraints on the SME coefficients. The four odd harmonics only contain cross terms of the SME coefficients, but the five even harmonics ($n = 0, 2, 4$) contain terms dependent only on individual coefficients. They can therefore be used to obtain measurements of the five unbounded components of $g_{\mu\kappa\lambda}$.

References

1. V.A. Kostelecký and N. Russell, *Data Tables for Lorentz and CPT Violation,* 2025 edition, arXiv:0801.0287v18.
2. P. Mohanmurthy, A. Narayan, and D. Dutta, *Mod. Phys. Lett. A* **31**, 1650220 (2016).
3. J.-P. Bocquet *et al.*, *Phys. Rev. Lett.* **104**, 241601 (2010).
4. V. Gharibyan, *Phys. Lett. B* **611**, 231 (2005).
5. A. Zec *et al.*, *Phys. Rev. C* **109**, 024323 (2024).
6. A. Narayan *et al.*, *Phys. Rev. X* **6**, 011013 (2016).
7. T. Allison *et al.*, *Nucl. Instrum. Meth. A* **781**, 105 (2015).

Chapter 30

NOPTREx — Neutron Optics Parity & Time-Reversal Experiment

Sepehr Samiei and W.M. Snow

Physics Department, Indiana University Bloomington, IN 47405, USA
Indiana University Center for Spacetime Symmetries, Bloomington, IN 47408, USA
Center for Exploration of Energy and Matter, 2401 Milo B. Sampson Lane, Bloomington, IN 47408, USA

We report on an ongoing program at the ANNRI beamline (J-PARC) to close the long-standing $140 < A < 180$ mass-window gap in parity-violating (P-odd) epithermal-neutron resonances. By discovering and characterizing new p-wave resonances, we will extract the mean-square weak matrix element $\langle |M_w|^2 \rangle$ in heavy nuclei and confront statistical models of P violation. Complementary searches for P-odd/T-odd and P-even/T-odd nucleon–nucleon interactions remain future objectives and are mentioned only in passing here.

1. Introduction

NOPTREx aims to exploit the natural "dynamical enhancement" that occurs when epithermal neutrons linger inside narrow p-wave compound-nuclear states. Measurements by the TRIPLE collaboration revealed spectacular P-odd asymmetries ($A_{\rm PV} \approx 10^{-3}$) in a handful of nuclei such as ^{139}La, but almost no data exist for $140 < A < 180$. Filling this gap is essential for a statistically robust determination of the weak nucleon–nucleon mixing strength in heavy nuclei.[1]

2. Physics Goals

- Discover new P-odd resonances in the $140 < A < 180$ heavy-nucleus region.
- Quantify the mean-square weak matrix element $\langle |M_w|^2 \rangle$ via statistical spectroscopy.
- Compare results with random-matrix predictions for compound-nuclear PV.

Complementary objectives—searches for P-odd/T-odd and P-even/T-odd NN interactions—are deferred to future phases.

3. Experimental Method (ANNRI, J-PARC)

Epithermal neutrons (0.1–10 eV) are produced by the JSNS spallation source and transported to the ANNRI beamline. A high-efficiency ^{3}He spin-filter provides longitudinal neutron polarization ($P \approx 0.97$) with spin-flip efficiency $\epsilon_{SF} \geq 0.999$. The target stack mounts candidate isotopes whose s- and p-wave strength functions S_0, S_1 satisfy $0.3 \lesssim S_1/S_0 \lesssim 10$, maximizing PV sensitivity. Transmission spectra are recorded in time-of-flight mode; the helicity-dependent total cross-section difference gives $A_{PV}(E)$.

4. Preliminary Results

The well-established p-wave resonance in ^{139}La at $E_r \approx 0.75$ eV exhibits a parity-violating asymmetry of $(9.5 \pm 0.2_{stat} \pm 0.3_{syst}) \times 10^{-4}$, providing an excellent in-beam calibration for ANNRI analysis tools.[2] Systematic studies of magnetic-field drifts and depolarization are ongoing. No newly surveyed resonance in the $140 < A < 180$ window exceeds $|A_{PV}| \approx 2 \times 10^{-5}$ within statistical significance.[3]

5. Outlook

Beam time for a dedicated high-statistics run has been granted under J-PARC Long-Term Proposal #R17. Combined with upgraded DAQ and improved neutron flux diagnostics, we anticipate a ten-fold increase in sensitivity to $\langle |M_w|^2 \rangle$ across the $140 < A < 180$ window within two years.

References

1. D. Bowman *et al.*, *Ann. Rev. Nucl. Part. Sci.* **43**, 829 (1993).
2. A. Mitchell and A. Rich, *Phys. Rev. C* **55**, MR23 (1997).
3. S. Samiei *et al.*, *First ANNRI Survey of Heavy-Nucleus PV,* to appear.

Chapter 31

Neural-Assisted Inference of High-Dimensional Symmetry-Breaking Coefficients

Jacob R. Schaefer,[*] Malachy Bloom,[*] Jolien Creighton,[†] Anarya Ray,[‡] and Jay D. Tasson[*]

[*]*Physics and Astronomy Department, Carleton College, Northfield, MN 55057, USA*

[†]*Department of Physics, University of Wisconsin-Milwaukee, Milwaukee, WI 53201, USA*

[‡]*CIERA, Northwestern University, Evanston, IL 60201, USA*

We summarize an end-to-end Bayesian framework for simultaneously constraining the 25 nonbirefringent mass-dimension-six Lorentz-invariance-violating coefficients of the gravitational Standard-Model Extension using observations from the third Gravitational-Wave Transient Catalog. We use a hybrid strategy, combining singular-value decomposition with an adaptive neural Markov-chain Monte Carlo algorithm based on continuous normalizing flows. Applied to 31 compact-binary coalescence events, the method yields significant speed-ups while preserving full Bayesian fidelity, making joint inference feasible for high-dimensional parameter spaces, thereby demonstrating the potential of neural-assisted approaches in high-dimensional tests of fundamental symmetries.

1. Introduction

Precision tests of Lorentz symmetry with gravitational waves (GWs) have evolved into a powerful probe of quantum gravity theories.[1] In the gravitational Standard-Model Extension (SME), Lorentz-invariance-violating

(LIV) coefficients can generate nonbirefringent, nondispersive time-of-flight signals[2,3]; nonbirefringent dispersive signals[4,5]; and combinations of birefringent and dispersive effects.[6] The leading nonbirefringent dispersive signals arise from 25 independent components of a rank-8 coefficient tensor associated with a mass dimension $d = 6$ operator in the action. Jointly constraining this 25-dimensional parameter space in a coefficient separation approach[7] is statistically challenging because the likelihood occupies a relatively small volume compared to that of the initial parameter space. This mismatch between the prior and posterior means that the prior–posterior Kullback–Leibler (KL) divergence,[8] a measure of information gain required to move from a prior distribution to the posterior distribution, can be large, indicating a more challenging sampling problem. In addition, the posterior potentially has a nontrivial structure and is informed by a set of GW events only slightly larger than the set of Lorentz-violating degrees of freedom to be constrained. Hence, we consider this limit of the SME as an example of the methods we present here.

Conventional nested-sampling or Markov-chain Monte Carlo (MCMC) approaches scale poorly with dimension. Here we outline a hierarchical strategy that (i) utilizes singular-value decomposition (SVD) to obtain the Gaussian-error posterior for the SME coefficients, identifying well-measured directions in parameter space, then (ii) trains a continuous normalizing flow (CNF) to learn an efficient, invertible reparameterization that drastically reduces the prior–posterior KL divergence before (iii) performing a kernel-density-estimation (KDE)-accelerated joint inference. The remainder of this proceedings contribution addresses these steps in turn.

2. Theory

The LIV corrections to linearized general relativity at mass dimension $d = 6$ in the gravitational sector of the SME arise via the Lagrange density[1]

$$\mathcal{L} \supset \frac{1}{4} h_{\mu\nu} K^{(6)\mu\nu\rho\sigma\epsilon_1\epsilon_2\epsilon_3\epsilon_4} \partial_{\epsilon_1}\partial_{\epsilon_2}\partial_{\epsilon_3}\partial_{\epsilon_4} h_{\rho\sigma}, \tag{1}$$

where $h_{\mu\nu}$ is the metric perturbation and $K^{(6)\mu\nu\rho\sigma\epsilon_1\epsilon_2\epsilon_3\epsilon_4}$ is a set of coefficients. This adds a fourth-order term in momentum space, eliciting a GW dispersion relation, which, with a modified interpretation, can be mapped onto the phenomenological model[9]

$$E^2 = p^2 + A_4 p^4, \quad \omega = |\mathbf{p}| \left(1 + \frac{1}{2} A_4 \omega^2\right), \tag{2}$$

where E is the energy, p is the momentum, A_4 is a phenomenological parameter containing GW dispersion information at $d = 6$, and ω is the frequency. The parameter A_4 is now to be interpreted as event dependent in this mapping, as it depends on sky direction $\hat{n}$ via the spherical-harmonic decomposition:

$$-\frac{1}{2}A_4 = \sum_{\ell=0}^{4} \sum_{m=-\ell}^{\ell} k^{(6)}_{(I)\ell m} Y_{\ell m}(\hat{n}), \tag{3}$$

where $k^{(6)}_{(I)\ell m}$ are the SME coefficients in a spherical basis.[1] This is used to turn existing A_4 posteriors obtained for each of a set of GW events into constraints on SME coefficients.[5] For example, 31 GW Transient Catalog Three[10] events can be used to constrain all 25 SME coefficients.

3. Singular Value Decomposition

Singular value decomposition is an extremely efficient method for solving an overconstrained system of linear equations. For $N_{\text{ev}} = 31$ events and $N_{\text{SME}} = 25$ SME coefficients, we can write the system of interest to us as

$$\boldsymbol{A}_4 = \boldsymbol{M}\hat{\boldsymbol{k}}, \tag{4}$$

where $\boldsymbol{A}_4$ is a vector of A_4 measurements for each of our events, $\hat{\boldsymbol{k}}$ is a vector of the 25 SME coefficients we seek to estimate, and the matrix $\boldsymbol{M}$ has entries $M_{ij} = [Y_{\ell m}]_j(\hat{n}_i)$ evaluated at each event's sky location. Singular value decomposition decomposes $\boldsymbol{M}$ into a diagonal matrix $\boldsymbol{\Sigma}$ with entries known as singular values and two unitary matrices $\boldsymbol{U}$ and $\boldsymbol{V}$:

$$\boldsymbol{M} = \boldsymbol{U}\,\boldsymbol{\Sigma}\,\boldsymbol{V}^{\mathsf{T}}, \tag{5}$$

which is used to identify orthogonal directions in the coefficient space. Inverting the linear system yields a point estimate of the SME coefficients[3]:

$$\hat{\boldsymbol{k}} = \boldsymbol{V}\,\boldsymbol{\Sigma}^{-1}\,\boldsymbol{U}^{\mathsf{T}}\boldsymbol{A}_4. \tag{6}$$

However, it only gives a pointwise estimate; it is a frequentist method that does not give true Bayesian credible regions. It relies on Gaussian-error assumptions and thus cannot capture nonGaussian or multimodal uncertainty structure. Large singular values correspond to well-constrained combinations; small ones expose degeneracies. The SVD also provides an orthogonal reparameterization of the coefficient space used later to precondition the CNF.

4. Hierarchical Kernel-Density Estimation

A KDE approach to estimating SME coefficients[3] is a fully Bayesian inference method that smooths each event's posterior on the effective dispersion parameter A_4 into a continuous likelihood $p_{KDE}(A_4)$. Simultaneous inference can then be performed by building a smooth joint posterior over all events using a Gaussian kernel by analogy with the approach provided by Eq. (18) of Ref. 3:

$$p(\boldsymbol{k}|d) \propto \prod_i \int p_{KDE,i}(A_4,\theta,\phi)\delta(A_4 - A_4(\theta,\phi,\boldsymbol{k}))dA_4 d\theta d\phi. \tag{7}$$

The product runs over our chosen events, (θ, ϕ) are sky-position coordinates over which we marginalize, and $p_{KDE,i}(A_4, \theta, \phi)$ is the density of the three random variables. However, this scales very poorly in high dimensions. We introduce CNF to circumvent this issue.

5. Continuous Normalizing Flows

A continuous normalizing flow[11] is a neural network that defines a family of diffeomorphisms $\{F_t : \mathbb{R}^{25} \to \mathbb{R}^{25}\}_{t\in[0,1]}$ satisfying

$$\partial_t F_t(\boldsymbol{z}) = \boldsymbol{u}_t\big(F_t(\boldsymbol{z})\big), \quad F_0 = \text{identity}. \tag{8}$$

Here $\boldsymbol{u}_t$ is a t-dependent velocity field whose functional form is parameterized by a neural network, with parameters learned from training data to produce a diffeomorphism that optimally maps the base distribution to the target distribution. The push-forward (denoted $[\cdot]_\sharp$) of a base distribution p_0 with respect to F_t is

$$p_t = [F_t]_\sharp p_0 = p_0\big(F_t^{-1}(\boldsymbol{z})\big)\big|\det J(F_t^{-1})\big|, \tag{9}$$

where $J(F_t^{-1})$ is the Jacobian of the flow transformation. The diffeomorphisms thus define a continuous set of t-dependent distributions $\{p_t\}_{t\in[0,1]}$, the final member $p_{t=1}$ being our target distribution.

If the random variable $\boldsymbol{z}_0 \sim p_0$ and $\boldsymbol{z}_t = F_t(\boldsymbol{z}_0)$, the induced density path p_t satisfies the continuity equation

$$\partial_t p_t + \partial_i\big(u_t^i\, p_t\big) = 0, \tag{10}$$

and integrating yields the above push-forward equation. Automatic differentiation through the network provides the exact divergence term $\partial_i u_t^i$ needed to calculate the Jacobian determinant, so reparameterizing the prior with a CNF leaves Bayesian evidence unchanged, ensuring the transformation is Bayesian-consistent.

We train the flow using flow-matching loss[12] to align the CNF's push-forward with an SVD estimate of the posterior. The initially trained flow is used as a prior transformation for samples proposed in an MCMC sampler with the KDE-built likelihood in Eq. (7). In addition, we adaptively update the network parameters throughout the sampling process. This initializes sampling in areas of high likelihood, resulting in convergence rates that are orders of magnitude faster. This reduces the KL divergence between prior and posterior distributions, which essentially counts the "information gain" needed to go from prior to posterior. Intuitively, a very large KL divergence between prior and posterior means the data drastically constrains the parameters (the posterior is much narrower than the prior), which typically implies more work for the sampler to locate and characterize that tiny high-likelihood volume. By using the CNF transform, we effectively reduce the KL divergence between the sampling distribution and the posterior. This permits a fully Bayesian approach to the simultaneous treatment of all 25 coefficients in this limit of the SME, outlining an approach with the potential to find numerous applications in high mass dimension SME searches.

Acknowledgments

The authors are grateful for computational resources provided by LIGO Laboratory and supported by National Science Foundation Grants PHY-0757058 and PHY-0823459.

References

1. V.A. Kostelecký and M. Mewes, *Phys. Lett. B* **757**, 510 (2016); M. Mewes, *Phys. Rev. D* **91**, 104062 (2019).
2. B.P. Abbott *et al.* (Virgo, Fermi-GBM, INTEGRAL, LIGO Scientific Collaboration), *Astrophys. J.* **848**, L13 (2017); X. Liu *et al.*, *Phys. Rev. D* **102**, 024028 (2020); N. Cornish, D. Blas, and G. Nardini, *Phys. Rev. Lett.* **119**, 161102 (2017).
3. A. Ray, P. Fan, V.F. He, M. Bloom, S.M. Yang, J.D. Tasson, and J.D.E. Creighton, *Phys. Rev. D* **110**, 122001 (2024).
4. C. Gong *et al.*, *Phys. Rev. D* **107**, 124015 (2023).
5. M. Bloom *et al.*, in preparation.
6. L. Haegel *et al.*, *Phys. Rev. D* **107**, 064031 (2023); R. Niu, T. Zhu, and W. Zhao, *J. Cosmol. Astropart. Phys.* **12**, 11 (2022).
7. N.A. Flowers, C. Goodge, and J.D. Tasson, *Phys. Rev. Lett.* **119**, 201101 (2017).

8. S. Kullback and R.A. Leibler, *Ann. Math. Stat.* **22**, 79 (1951).
9. S. Mirshekari, N. Yunes, and C.M. Will, *Phys. Rev. D* **85**, 024041 (2012).
10. R. Abbott *et al.* (LIGO Scientific, Virgo, and KAGRA Collaboration), *Phys. Rev. D* **112**, 084080 (2025).
11. R.T.Q. Chen, Y. Rubanova, J. Bettencourt, and D. Duvenaud, arXiv: 1806.07366.
12. Y. Lipman, R.T.Q. Chen, H. Ben-Hamu, M. Nickel, and M. Le, arXiv: 2210.02747.

Chapter 32

Using Slow-Neutron Polarimetry to Search for Exotic Spin-Dependent Fifth Forces

M. Van Meter,[*,†,‡] **K.N. Lopez,**[*,†,‡] **K. Li,**[*,†,‡]
T. Mulkey,[§] **and W.M. Snow,**[*,†,‡]

[*]*Physics Department, Indiana University, Bloomington, IN 47405, USA*
[†]*Indiana University Center for Spacetime Symmetries, Bloomington, IN 47408, USA*
[‡]*Center for Exploration of Energy and Matter, 2401 Milo B. Sampson Lane, Bloomington, IN 47408, USA*
[§]*Georgia State University, Atlanta, GA 30303, USA*

Through the use of polarized neutron imaging, our collaboration searches for exotic spin-spin interactions between neutrons and electrons. Using slow neutron polarimetry, we search for the rotation of the neutron's spin from its interaction with polarized electrons within a ferrimagnetic sample held at the compensation temperature in order to suppress its internal magnetic field. A positive signal could indicate beyond-the-Standard-Model physics. Our work sets the most stringent upper bound on the product of exotic axial neutron–electron couplings over a very broad range of spin-1 exchange-boson masses.[1]

1. Introduction

Possible beyond-the-Standard-Model physics could be realized through the mediation of weakly interacting light bosons. An exchange boson with a mass of 10^{-5} eV possesses a Compton wavelength (and therefore an interaction range) on the scale of 0.2 cm.[2] In the nonrelativistic limit, such long-range interactions can be classified in terms of scalar combinations of their spins and momenta. In our case the potential of most immediate interest is

$$V_2 = f_2^{eN} \frac{\hbar c}{4\pi} \hat{\sigma}_1 \cdot \hat{\sigma}_2 \left(\frac{1}{r}\right) e^{-r/\lambda_c}, \tag{1}$$

where $\sigma_{1,2}$ represent the spin of the neutron and electron, respectively, λ_c is the range of the exotic interaction, and r is the distance between the neutron and the electron. Our measurement can place an upper bound on the electron–neutron axial coupling constant product $f(0,0) = -g_A^e g_A^N$.[3]

By adding these two-body interactions coherently over the large number of atoms in a macroscopic target medium, we can greatly increase the sensitivity of this search.[2] However, for a spin-dependent interaction of the type we seek, there is always the danger of a systematic error coming from magnetic interactions. We have succeeded in finding a system where this background effect can be greatly neutralized. We employ a ferrimagnetic sample held at its compensation temperature to act as our source of polarized electrons. Ferrimagnets contain distinct sublattices of atoms with magnetic moments of differing magnitudes arranged in an antiferromagnetic pattern. Because of this, a ferrimagnet still, in general, possesses a net magnetization despite its antiferromagnetic coupling. Due to the differing temperature dependence of the magnetization of the different sublattices, many ferrimagnets possess a temperature known as the magnetic compensation temperature (T_{comp}) where the magnetic moment of the ferrimagnetic sample goes to zero at the unit-cell level. However, since the gyromagnetic ratios of the two atoms may differ, the unit cell can still possess a total spin angular momentum $\vec{s}$ at T_{comp}.[4] We utilize the compound terbium iron garnet ($Tb_3Fe_5O_{12}$, or TbIG for short) with Fe and Tb acting as our differing sublattice species. Unlike Fe, whose moment depends mainly on the spin of the electrons, the magnetic moment of Tb is a result of both the spin and orbital motion of the inner 4f electrons.[4] We can therefore realize both a net electron polarization and also a zero net magnetization at the unit-cell level when our sample is held at T_{comp}.

Polarized neutron spin rotation acts as the probe in our search for these exotic interactions. One can manipulate and analyze the neutron's spin with ease, and the interactions of the neutron in the medium are mostly coherent, thereby allowing the rotation angle effect from the exotic interaction to accumulate as the neutrons move deeper into the medium. The spin state of transversely polarized neutrons relative to the momentum can be written as a coherent superposition of positive and negative helicity spin states. In the presence of the exotic interaction described above, these two states will accumulate phases equal in magnitude and opposite in sign, as seen with Eq. (1).[5] This results in a rotation of the neutron's spin about the neutron momentum. The resulting rotation angle can be extracted through neutron polarimetry techniques.

2. Experimental Setup and Analysis

For our source of free neutrons, we employed the CG-1D beamline at the High Flux Isotope Reactor at Oak Ridge National Laboratory. Utilizing a supermirror polarizer and spin transport elements, the incoming neutrons are kept in a vertically polarized state as they approach the sample region. Because a precession of the neutron's spin due to an exotic force is indistinguishable from Larmor precession due to ambient magnetic-field backgrounds, the sample region is held within a cylindrical μ-metal magnetic shield. The rotation angle from residual magnetic fields is subtracted by rotating the sample by 180° and then taking the difference in rotation angles. The 180° rotation of the sample reverses the sign of the electron polarization, whose direction is tied to the material. A block diagram of the beamline components showing how the neutron's spin is guided is shown in Fig. 1.[4]

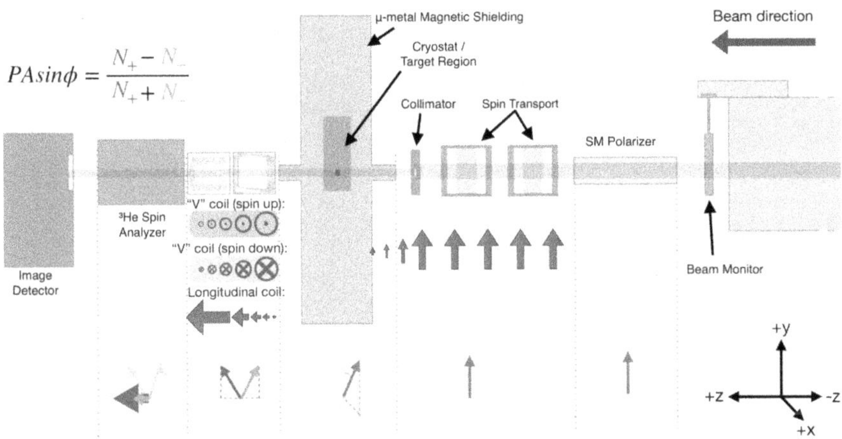

Fig. 1. The polarimetry components and manipulation of the neutron's spin on the CG-1D beamline. The bottom row of vectors indicate the direction of the neutron's spin and the arrows above show the direction of the magnetic fields in each section of the beamline.

The sample was held within an aluminum cryostat maintained at a rough vacuum for temperature stability, with the sample thermally connected to a cooled copper block and maintained at T_{comp} through our PID temperature control system.[4] During operation, we were able to maintain a vacuum of roughly 19 mTorr with a temperature stability of 100 mK when held at T_{comp} in the 2023 run of the experiment. Through our previous neutron spin echo measurements, we found T_{comp} to be approximately 250.3 ± 0.1 K.[4] A more detailed description of the experimental setup is described in Ref. 4.

Once exiting the sample region, we utilize spin transport coils to adiabatically rotate the neutron's spin into the longitudinal plane so that it may be filtered by a longitudinally polarized ^{3}He neutron polarization analyzer and then captured by the imaging detector.[4]

By reversing the direction of the current in our "V" coil, we are able to extract the normalized asymmetry between the two spin states, which in turn gives the net spin rotation angle of the neutrons through the equation

$$PA\sin(\phi) = \frac{N_+ - N_-}{N_+ + N_-} \tag{2}$$

where $N_\pm$ are the number of neutrons the imaging detector collects for each spin state, P is the neutron polarization (≈ 0.95), A is the analyzing power of the neutron polarization analyzer (≈ 0.86), and ϕ is the rotation of the neutron's spin.[4] The neutron counts $N_\pm$ are extracted directly from the images collected from the detector.

3. Measurement Result

From our 2023 data, we measured an asymmetry of $1.99 \pm 9.62 \times 10^{-5}$. This measurement was made after 34 separate runs where we produced a total of 535 images. Each image was taken with 2.5 minutes of exposure time in order to reduce possible temperature or ambient magnetic-field drift. Figure 2 shows our results.

Our results are in agreement with zero at the 10^{-5} level. This places a stringent upper bound on $f(0,0) = -g_A^e g_A^N$. We plan to reduce the systematic uncertainty of our experiment through developing a more stable temperature control system and employing more μ-metal shielding for a stronger suppression of the ambient magnetic field.

We still have a lot of potential in probing for exotic forces with our novel apparatus as shown by our 2026 publication improving the upper bounds

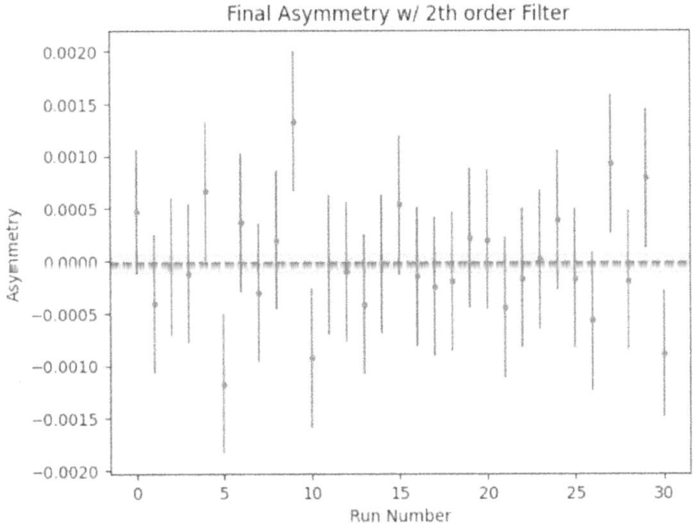

Fig. 2. Measured asymmetries after a second-order filter to remove ambient magnetic-field effects.

on $g_A^e g_A^N$ in the $10^{-8} \leq \lambda_c \leq 10^{-2}$ m range by several orders of magnitude[1] and another data set taken during November of 2025 that is currently under analysis focusing on probing different exotic interactions by changing the direction of the electron polarization. More details will be given on this data set as well as upcoming experimental runs in future publications.

References

1. T. Mulkey *et al. Phys. Rev. Lett.* 136, 071801 (2026).
2. J.E. Moody and F. Wilczek, *Phys. Rev. D* **30**, 130 (1984).
3. B.A. Dobrescu and I. Mocioiu, *JHEP* **11**, 005 (2006).
4. C.D. Hughes *et al.* J. *Magn. Magn. Mater.* **629**, 173273 (2025).
5. V.F. Sears, *Neutron Optics: An Introduction to the Theory of Neutron Optical Phenomena and their Applications,* Oxford University Press, Oxford, 1987.

Chapter 33

Do Anomalies Break Momentum-Routing Invariance?

A.R. Vieira

Instituto de Ciências Agrárias, Exatas e Biológicas de Iturama - ICAEBI, Universidade Federal do Triângulo Mineiro - UFTM, Iturama, Minas Gerais 38280-000, Brazil

The diagrammatic computation of the chiral anomaly is associated with momentum-routing invariance breaking. This happens because the momentum routing in the internal lines of a loop diagram is chosen such that the gauge Ward identities hold and the chiral Ward identity is broken by the finite term measured in the pion decay into two photons. Since the latter is observable, it seems that there is a preferred momentum routing set by experiments. However, it is shown in this work that the chiral anomaly is momentum-routing invariant. This idea is specially important for situations in which there are no experiments yet to decide on the momentum routing, like in supersymmetric theories and in frameworks with CPT and Lorentz violation. Therefore, we resort to momentum-routing invariance to find out what symmetry is broken in a nonminimal CPT- and Lorentz-violating version of quantum electrodynamics.

1. Introduction

Although symmetries are one of the main aspects of classical field theories, we know that is not possible to maintain all of them at the quantum level. The name anomaly is given to a breaking of a classical symmetry caused by quantum corrections. From the physical point of view, we are interested in investigating if such anomalies are spurious or indeed physical. In the former case, it is caused by the regularization scheme, that usually breaks symmetries of theories, and restoring counter terms are required in the renormalization process. In the latter case, the anomaly can be measured in an experiment, like pion decay into two photons, which reveals that chiral symmetry no longer holds at the quantum level.[1]

Since its discovery,[1] the chiral anomaly was derived by several different approaches. The most common ones, usually found in textbooks, rely on the fact that we should correctly choose the momentum routing in the internal lines of the triangle diagrams such that the gauge Ward identities hold and the chiral Ward identity is broken by the term measured in pion decay into two photons. Therefore, it seems that there is a preferred momentum routing set by experiment. This might lead to the conclusion that momentum-routing invariance in Feynman diagrams no longer holds because it was necessary to choose a specific momentum routing in order to get the desired results in the Ward identities. However, the reciprocal is not true, i.e., anomalies do not imply momentum-routing invariance breaking. It is also possible to derive the correct result for the anomaly in a momentum-routing invariant way, as will be presented in this work.

2. Momentum-Routing Invariance of Feynman Diagrams

Feynman diagrams present the feature of being momentum-routing invariant. As a simple example, we can consider the one-loop process presented in Fig. 1, which is a quantum correction to electron scattering. As we see, the amplitude of this process should not depend on whether the momentum of the internal lines is labeled as k or $k + l$ because the cross section is an observable. However, this feature is not usually an issue because the most popular regularization scheme is dimensional regularization (DReg), and this scheme allows for shifts in the integrated momentum. In other words, DReg is momentum-routing invariant besides being gauge invariant.

In abelian and chiral abelian gauge field theories, there is an additional reason that leads to momentum-routing invariance. This invariance is obeyed if and only if gauge invariance is as well. As a simple example,

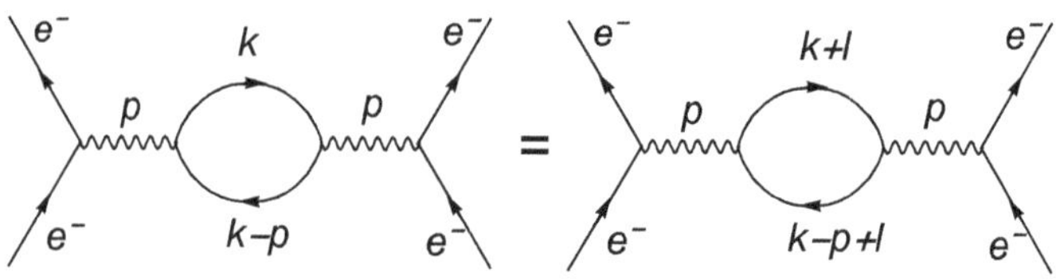

Fig. 1. Momentum-routing invariance of a QED process.

let us consider the gauge Ward identity obtained from the vacuum polarization tensor. When the amplitude $\Pi^{\mu\nu}(p)$ is contracted with an external momentum, we find that this gauge Ward identity is nothing but the difference between two tadpoles with different momentum routing each. This diagrammatic proof is independent of regularization, it can be simply achieved algebraically as below and it can be generalized to diagrams with more legs and loops:

$$\begin{aligned} p_\mu \Pi^{\mu\nu}(p) &= -e^2 \int_k \mathrm{Tr}\left[\not{p}\frac{1}{\not{k}-\not{p}-m}\gamma^\nu \frac{1}{\not{k}-m}\right] \\ &= -e^2 \int_k \mathrm{Tr}\left[\gamma^\nu \frac{1}{\not{k}-\not{p}-m}\right] + e^2 \int_k \mathrm{Tr}\left[\gamma^\nu \frac{1}{\not{k}-m}\right]. \end{aligned} \tag{1}$$

The gauge and momentum-routing invariance relation can be confirmed with regularization schemes that do not break gauge or momentum routing invariances.

3. Momentum-Routing Invariant Chiral-Anomaly Derivation

The computation of the chiral anomaly is performed for general momentum routings with implicit regularization.[2] In this approach, the regularization-dependent content resides in the surface terms. The corresponding amplitudes of the triangle diagrams are

$$\begin{aligned} T_{\mu\nu\alpha} = -i \int_k \mathrm{Tr}\left[\gamma_\mu \frac{i}{\not{k}+\not{k}_1-m}\gamma_\nu \frac{i}{\not{k}+\not{k}_2-m}\gamma_\alpha\gamma^5 \frac{i}{\not{k}+\not{k}_3-m}\right] \\ + (\mu \leftrightarrow \nu, p \leftrightarrow q). \end{aligned} \tag{2}$$

General momentum routings are constrained by energy-momentum conservation at each vertex: $k_2 - k_3 = p+q$, $k_1 - k_3 = p$, and $k_2 - k_1 = q$. This allows us to parameterize general momentum routings as $k_1 = \alpha p + (\beta-1)q$, $k_2 = \alpha p + \beta q$, and $k_3 = (\alpha-1)p + (\beta-1)q$, where α and β are real numbers. After computing the diagrams and applying the respective external momentum in order to obtain the Ward identities, we find the following results:

$$\begin{aligned} p_\mu T^{\mu\nu\alpha} &= -4iv_0(\alpha-\beta-1)\epsilon^{\alpha\nu\beta\lambda}p_\beta q_\lambda, \\ q_\nu T^{\mu\nu\alpha} &= 4iv_0(\alpha-\beta-1)\epsilon^{\alpha\mu\beta\lambda}p_\beta q_\lambda, \\ (p+q)_\alpha T^{\mu\nu\alpha} &= 2mT_5^{\mu\nu} + 8iv_0(\alpha-\beta-1)\epsilon^{\mu\nu\beta\lambda}p_\beta q_\lambda - \tfrac{1}{2\pi^2}\epsilon^{\mu\nu\beta\lambda}p_\beta q_\lambda, \end{aligned} \tag{3}$$

where $T_5^{\mu\nu}$ is the usual vector-vector-pseudoscalar triangle.

It is easy to see in Eq. (3) that if the surface term υ_0 is zero, we get the correct result of the chiral anomaly $-\frac{1}{2\pi^2}\epsilon^{\mu\nu\beta\lambda}p_\beta q_\lambda$ and gauge symmetry holds in the first two equations. Besides, this result is momentum-routing invariant since α and β can be any real numbers for a null surface term.

4. The Chiral Anomaly in a $d = 5$ Lorentz-Violating QED

We can resort to momentum-routing invariance to find out what symmetry is broken in situations where there are no experiments yet to fix the momentum-routing ambiguity.[2] This would be the case in supersymmetric theories or frameworks with CPT and Lorentz violation. We consider the mass dimension $d = 5$ operator in the fermion sector $-\frac{1}{2}(a_F^{(5)})^{\mu\alpha\beta}\bar{\psi}\gamma_\mu F_{\alpha\beta}\psi$ in the Lagrangian $\mathcal{L}_{\psi F}^{(5)}$ of Table I in Ref. [3], where $(a_F^{(5)})^{\mu\alpha\beta}$ is a set of CPT- and Lorentz-violating coefficients. It can be rewritten as $-(a_F^{(5)})^{\mu\alpha\beta}\bar{\psi}\gamma_\mu\partial_\alpha A_\beta\psi$ due to antisymmetry in α, β.

The triangle diagrams are depicted in Fig. 2. As in the usual case, this contribution must be summed with the crossed diagrams. Its corresponding amplitude can be written as

$$T_{\mu\nu\alpha}^{\mathrm{LV}} = -i\int_k \mathrm{Tr}\left[\gamma_\mu \frac{i}{\not{k}+\not{k}_1-m}(a_F^{(5)})_{\zeta\lambda\nu}\gamma^\zeta q^\lambda \frac{i}{\not{k}+\not{k}_2-m}\gamma_\alpha\gamma^5\frac{i}{\not{k}+\not{k}_3-m}\right]$$
$$-i\int_k \mathrm{Tr}\left[(a_F^{(5)})_\mu^{\zeta\lambda}\gamma_\zeta p_\lambda \frac{i}{\not{k}+\not{k}_1-m}\gamma_\nu\frac{i}{\not{k}+\not{k}_2-m}\gamma_\alpha\gamma^5\frac{i}{\not{k}+\not{k}_3-m}\right]$$
$$+(\mu\leftrightarrow\nu, p\leftrightarrow q). \quad (4)$$

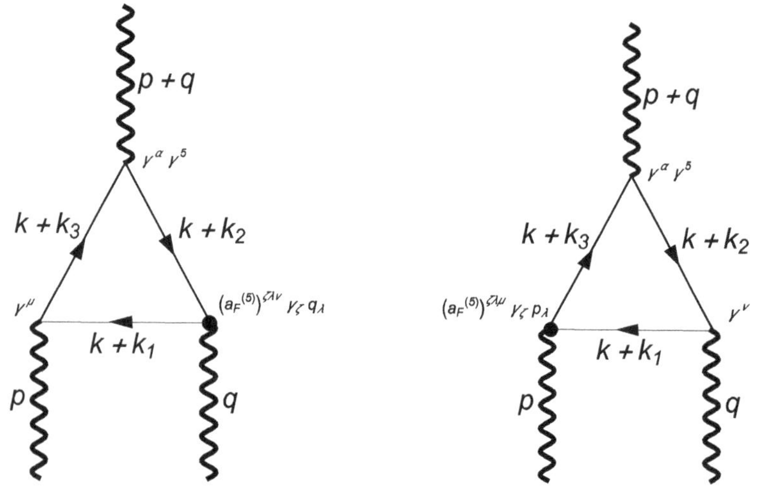

Fig. 2. Triangle diagrams in a QED with nonminimal Lorentz violation.

The parameterization for the arbitrary routing of k_i is the same as in the previous section. Now, if we compute the Ward identities we find, as expected, that gauge symmetry is broken by the arbitrary surface term υ_0, as presented below for the massless limit

$$\begin{aligned} p_\mu T^{LV\,\mu\nu\alpha} &= -\,8i\beta\upsilon_0(a_F^{(5)})_\varsigma{}^{q\nu}\epsilon^{q\alpha\zeta p}, \\ q_\nu T^{LV\,\mu\nu\alpha} &= 8i(1-\alpha)\upsilon_0(a_F^{(5)})_\varsigma{}^{p\mu}\epsilon^{q\alpha\zeta p}, \\ (p_\alpha+q_\alpha)T^{LV\,\mu\nu\alpha} &= 8i(\alpha-\beta-1)\upsilon_0\left[(a_F^{(5)})_\varsigma{}^{q\nu}\epsilon^{q\zeta\mu p}-(a_F^{(5)})_\varsigma{}^{p\mu}\epsilon^{q\zeta\nu p}\right] \\ &\quad -\frac{1}{2\pi^2}\left[(a_F^{(5)})_\varsigma{}^{q\nu}\epsilon^{q\zeta\mu p}+(a_F^{(5)})_\varsigma{}^{p\mu}\epsilon^{p\zeta\nu q}\right], \end{aligned} \tag{5}$$

where $(a_F^{(5)})^{\alpha p\mu} \equiv (a_F^{(5)})^{\alpha\lambda\mu}p_\lambda$ and $\epsilon^{\alpha\lambda\nu p} \equiv \epsilon^{\alpha\lambda\nu\sigma}p_\sigma$.

In this case, there is no particle process, such as pion decay, where the breaking of chiral symmetry provided by Lorentz violation at tree level is measured. So, the ambiguity in the momentum routing can not be solved by means of an experiment. However, since we showed in the previous section that momentum-routing invariance is not broken in the usual chiral anomaly, we can resort to this invariance in Eq. (5). We see that the results of the Ward identities are momentum-routing invariant if the surface term υ_0 is null. As a consequence, we have gauge symmetry and a chiral anomaly induced by CPT and Lorentz violation at the quantum level. It is noteworthy that some choices of momentum routing such as $\alpha = 0$ and $\beta = -1$ are compatible with the Ward identities of the Lorentz-invariant situation measured by experiments, i.e., gauge symmetry holding with no chiral symmetry, as we can see from Eq. (3). Nevertheless, this choice of momentum routing breaks both gauge and chiral symmetry in the Lorentz-violating situation as we can see in the Ward identities (5). We can instead alternatively choose a null value of the surface term, so that the Ward identities of both cases are compatible with momentum-routing invariance. In this case, we would have in both situations gauge symmetry and the breaking of chiral symmetry.

References

1. J.S. Bell and R. Jackiw, *Nuovo Cim. A* **60**, 47–61 (1969); S.L. Adler, *Phys. Rev.* **177** 2426 (1969).
2. A.R. Vieira, *Ann. Phys.* **478**, 170025 (2025).
3. V.A. Kostelecký and Z. Li, *Phys. Rev. D* **99**, 056016 (2019).

Chapter 34

Nonlinear ODEs in Bumblebee Gravity

Dario T. Walter-Cardona and Quentin G. Bailey

Physics and Astronomy Department,
Embry-Riddle Aeronautical University,
Prescott, AZ 86301, USA

In this work, we study a vector model of spontaneous spacetime-symmetry breaking coupled to gravity: the bumblebee model. The primary focus is on static spherically symmetric solutions. In this part of the work, we investigate the flat spacetime limit, which can be viewed as a modified electrostatics model with a nonlinear interaction term. We study the stability of classical solutions in the spherically symmetric case. This includes results with confluent hypergeometric functions, with interesting properties from QFT.[1]

1. The Bumblebee Model and the Flat-Spacetime Limit

Vast studies of signatures for spacetime-symmetry breaking in experiments and observations have been made in recent decades.[2–4] We start with the original bumblebee model, a vector model of spontaneous Lorentz violation with a potential driving the symmetry breaking and a Maxwell kinetic term,[5] in flat spacetime. The Lagrangian and field equations are as follows:

$$L = -\int d^3x \left(\tfrac{1}{4}B^{\mu\nu}B_{\mu\nu} + V\right), \tag{1}$$

$$F^{\mu\nu}{}_{,\mu} = 2V'B^{\nu}. \tag{2}$$

The potential is defined to be a function of the bumblebee field, $V = V(B^\mu B_\mu)$, and the prime is the derivative with respect to the argument. We then assume spherical symmetry and set the bumblebee field to be a timelike vector field, $B_\mu = (B_0(r), 0, 0, 0)$, such that $V = V(-B_0^2)$, and only a time component remains from the field equation in Eq. (2).

$$\nabla^2 B_0 = 2V' B_0. \tag{3}$$

We use a length scale Λ^{-1} associated with the minimum value of the potential to make our variables dimensionless, so that our final differential equation takes the form

$$\frac{d^2 f}{dx^2} + \frac{2}{x}\frac{df}{dx} = 2fV'(-f^2), \tag{4}$$

where $f = B_0/\Lambda$ and $x = \Lambda r$.

2. The Potentials

In this work, we look at three different forms of the potential function shown in Eq. (4). One model is the classic quadratic potential. The other two are based on the Kummer confluent hypergeometric functions. Sample potentials are shown in Fig. 1. The latter potentials were first proposed for

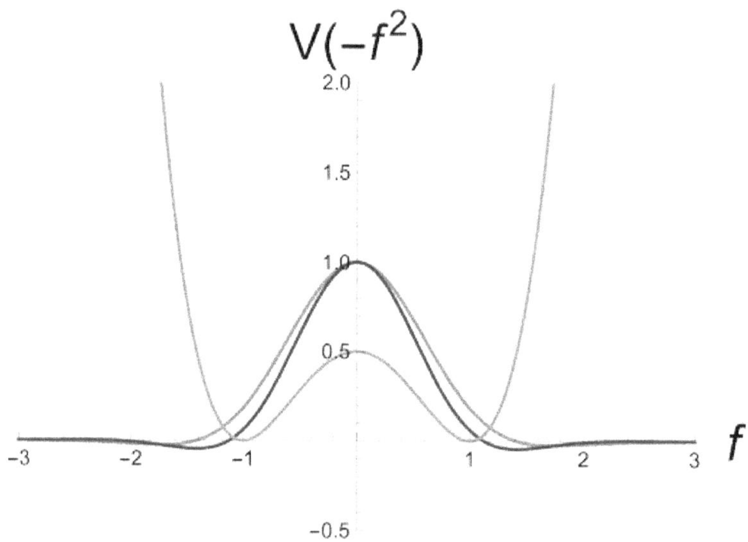

Fig. 1. Plots of the potential as functions of the time component of the vector potential. Orange takes the form $V(f) = \frac{1}{2}(1 - f^2)^2$, pink is the hypergeometric function $M(3, 2, -f^2)$, and magenta is the hypergeometric function $M(4, 2, -f^2)$.

the bumblebee model in 2005 in a paper by Altschul and Kostelecký. They showed that for an arbitrary potential for a vector field, asymptotically free theories can have renormalizable and stable quantum field theories for special choices of the potential.[6] The Kummer confluent hypergeometric functions are defined by:

$$M(a,b,z) = \sum_{n=0}^{\infty} \frac{a^{(n)} z^n}{b^{(n)} n!}, \tag{5}$$

while explicit closed-form expressions exist for an integer n when restricting the form to $M(n,2,z)$, where $z = -f^2$.

3. Analytical Stability Argument

Using the substitution $x = \frac{1}{u}$ in Eq. (4), we transform the differential equation so that the f' term is eliminated. We can then make generalized statements about the behavior of the $f(u)$ function for a given potential. Specifically for the quadratic potential and the two hypergeometric potentials, we summarize as follows:

$$u^4 \frac{d^2 f}{du^2} = 2f(1-f^2) \rightarrow \begin{Bmatrix} f_{uu} > 0: \ f < -1 \\ f_{uu} < 0: \ -1 < f < 0 \\ f_{uu} > 0: \ 0 < f < 1 \\ f_{uu} < 0: \ f > 1 \end{Bmatrix}, \tag{6}$$

$$u^4 \frac{d^2 f}{du^2} = e^{-f^2}\left(3 - f^2\right) f \rightarrow \begin{Bmatrix} f_{uu} > 0: \ f < -\sqrt{3} \\ f_{uu} < 0: \ -\sqrt{3} < f < 0 \\ f_{uu} > 0: \ 0 < f < \sqrt{3} \\ f_{uu} < 0: \ f > \sqrt{3} \end{Bmatrix}, \tag{7}$$

$$u^4 \frac{d^2 f}{du^2} = \frac{1}{6} e^{-f^2} f\left(f^4 - 8f^2 + 12\right) \rightarrow \begin{Bmatrix} f_{uu} < 0: \ f < -\sqrt{6} \\ f_{uu} > 0: \ -\sqrt{6} < f < -\sqrt{2} \\ f_{uu} < 0: \ -\sqrt{2} < f < 0 \\ f_{uu} > 0: \ 0 < f < \sqrt{2} \\ f_{uu} < 0: \ \sqrt{2} < f < \sqrt{6} \\ f_{uu} > 0: \ f > \sqrt{6} \end{Bmatrix}. \tag{8}$$

In the quadratic-potential case and the hypergeometric $n = 3$ case, whenever f is not at a point of stability, the concavity of $f(u)$ changes to move f towards one of the points of stability. This behavior occurs regardless of

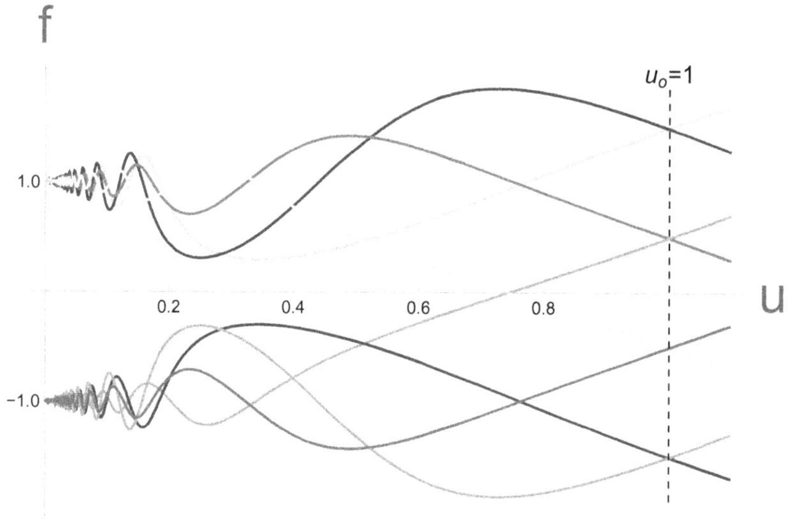

Fig. 2. The numerical solutions for the quadratic model, $V(-f^2) = \frac{1}{2}(1 - f^2)^2$.

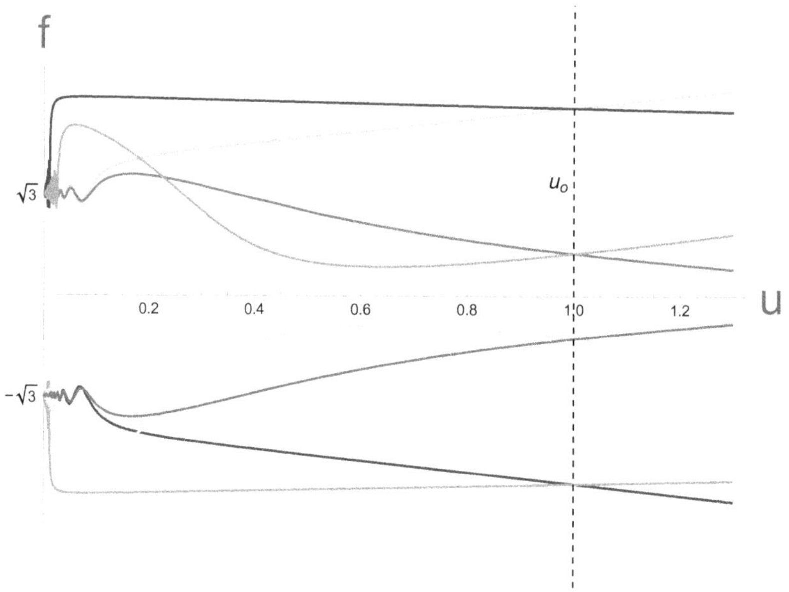

Fig. 3. The numerical solutions for the hypergeometric model, $V(-f^2) = M(3, 2, -f^2)$.

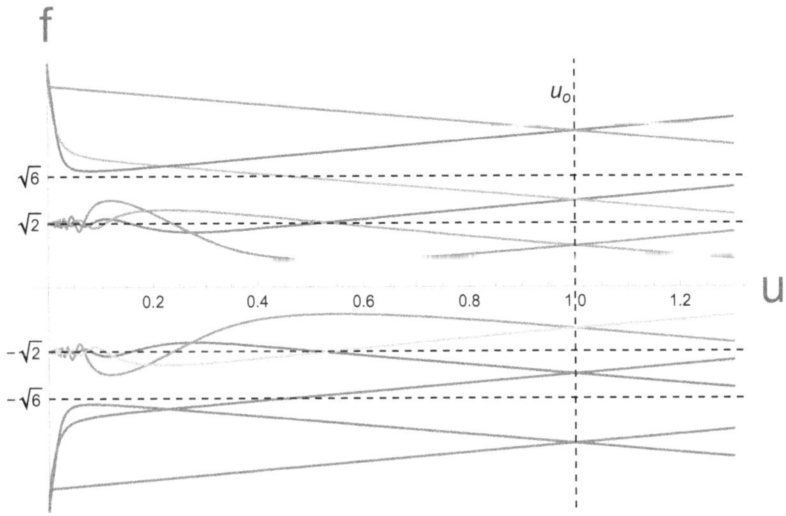

Fig. 4. The numerical solutions for the hypergeometric 4 model, $V(-f^2) = M(4, 2, -f^2)$.

the initial position or initial slope f_u, and the solution asymptotes to one of the two points of stability.

4. Numerical Solutions

The following graphs confirm our analytical stability argument. For decreasing values of u, f approaches its points of stability. This represents $x \to \infty$.

Figures 2 and 3 show the quadratic and hypergeometric potential respectively, which each have two points where the potential is at a minimum. Figure 4 however depicts numerical solutions for the hypergeometric potential, which has four points where the potential is at a minimum. As such, solutions are shown to diverge when they exceed values greater in magnitude than the outermost points.

Acknowledgments

Special thanks to Dr. Quentin Bailey for beginning this project and supporting the students on the team, to Dr. Edward Poon and Hailey Murray for laying the groundwork for a proper analysis of the flat-spacetime limit, and to the Undergraduate Research Institute (URI) and Dr. Anne Boettcher for supporting this project financially and providing professional and academic development resources. This work was supported in part by NSF grant no. 2308602 and the NASA Space Grant.

References

1. Q.G. Bailey, H.S. Murray, and D.T. Walter-Cardona, *Phys. Rev. D* **112**, 024069 (2025).
2. V.A. Kostelecký and N. Russell, *Data Tables for Lorentz and CPT Violation,* 2025 edition, arXiv:0801.0287v18.
3. T. Mariz, J.R. Nascimento, and A. Petrov, *Lorentz Symmetry Breaking—Classical and Quantum Aspects,* Springer Briefs in Physics, 2023, arXiv:2205.02594.
4. A. Addazi *et al.*, *Prog. Part. Nucl. Phys.* **125**, 103948 (2022).
5. V.A. Kostelecký and S. Samuel, *Phys. Rev. D* **40**, 1886 (1989).
6. B. Altschul and V.A. Kostelecký, *Phys. Lett. B* **628**, 106 (2005).

www.ingramcontent.com/pod-product-compliance
Lightning Source LLC
LaVergne TN
LVHW051052210926
848486LV00002B/79